Introduction to Earth and Planetary System Science

Naotatsu Shikazono

Introduction to Earth and Planetary System Science

New View of Earth, Planets and Humans

 Springer

Naotatsu Shikazono
Professor
Department of Applied Chemistry
Keio University
sikazono@applc.keio.ac.jp

CHIKYUU SISUTEMU KAGAKU NYUUMON
CHIKYUU WAKUSEI SISUTEMU KAGAKU NYUUMON
© 1992 Naotatsu SHIKAZONO
© 2009 Naotatsu SHIKAZONO
All rights reserved
Original Japanese edition published in 1992 and 2009 by University of Tokyo Press
English translation rights arranged with University of Tokyo Press
through Japan UNI Agency, Inc., Tokyo

ISBN 978-4-431-54057-1   e-ISBN 978-4-431-54058-8
DOI 10.1007/978-4-431-54058-8
Springer Tokyo Dordrecht Heidelberg London New York

Library of Congress Control Number: 2012933368

Springer is part of Springer Science+Business Media (www.springer.com)

Preface

More than 10 years have passed since the beginning of the twenty-first century, and problems associated with the global environment, resources, and the world economy have been distinctly realized. It is highly desirable to create long-term safety and a sustainable human society by solving these serious problems.

Is it possible to build a sustainable society? How can we construct such a society? Human society greatly influences the earth's environment, causing such problems as global warming, acid rain, and destruction of the ozone layer. Thus, it is inferred that long-term safety and a stable human society cannot be established without a scientific understanding of the earth's surface environment and the earth's interior as it influences the surface environment. We should consider what human society ought to do in the near future based on scientific understanding of interactions between humans and nature.

What are the nature–human interactions? The basic study that can give us the answer to these questions is earth system science, which has been developed significantly in the last 20 years. According to earth system science, the earth system consists of subsystems such as the atmosphere, hydrosphere, geosphere (lithosphere), biosphere, and humans, and earth system science clarifies the interactions among these subsystems. Matter and energy are circulating, and the total earth system has been irreversibly changing with time since the birth of the system. Earth system science can clarify the evolution of the earth system as well as the modern earth system.

Twenty years ago I described an outline of the earth system in *Introduction to the Earth System* (1992) (in Japanese). Following this publication, earth system science has developed considerably. For example, a great deal of information on solar planets, the relationship between the solar planets and earth, and the origin and evolution of earth and the solar system have been elucidated, clearly indicating that the earth system is open to external systems (the solar system; universe). In addition to the influence of the external (solar) system on the origin and evolution of the earth system, the materials, temperature distribution, and hot and cold plumes in the solid interior of the earth have been investigated by high-temperature and high-pressure experiments, three-dimensional seismic wave tomography, multi-element and multi-isotope analyses, and radiometric age dating of earth materials and solar system materials (e.g., Martian rocks). Further, global climate change from ancient times to the present and global material cycles (e.g., the global carbon cycle) have been studied by computer simulations and isotopic and chemical analyses. It can be said that earth system science is developing and is changing to earth and planetary system science. Thus, in this book, the intention is to present earth and planetary system science including topics that are not in my earlier publication, *Introduction to Earth System Science*. For example, plate tectonics and plume tectonics are briefly described in this book (Chap. 3) because the earth's surface environment where humans and organisms are living are greatly influenced by these tectonics. In addition, the relationship between humans and organisms, which has been extensively investigated (origin of life, evolution of biota, mass extinction, underground biosphere) has been included (Chap. 6). In the last chapter (Chap. 7), the relationship between the earth system and nature (earth and the planets) is considered. Earlier views regarding the earth and nature are summarized and compared with the views presented here (earth and planetary system sciences, earth and planet co-oriented human society).

This book has arisen mainly from several courses on Earth and Planetary System Science and Earth's Environmental and Resources Problems for undergraduate and graduate students at Keio University and also from many classes at other universities (The University of Tokyo, Gakushuin University, Nihon University, Hiroshima University, Yamaguchi University, Tokushima University, Kyoto University, Shizuoka University, Tsukuba University, Yamagata University, Tohoku University, and Akita University). In these courses during the last 30 years, I received many comments, questions, and responses from numerous undergraduate and graduate students. Discussions with them have allowed me to develop and clarify the ideas presented here, particularly the earth's environmental co-oriented society (Chap. 7).

In writing this book I am greatly indebted to many people in the Geology Department of The University of Tokyo, the Applied Chemistry Department of Keio University, the Geology Department of Tokyo Gakugei University, and the Department of Earth and Planetary Science of Harvard University.

I express my great appreciation for the late professors emeriti T. Tatsumi of The University of Tokyo, advisor for my Ph.D. thesis in 1974, and T. Watanabe of The University of Tokyo for teaching me economic geology (ore genesis) and isotope geochemistry. They showed that the integration of each discipline of earth sciences is necessary and very important in order to deeply understand the nature of the earth's environment and resources in relation to geological, geochemical, and geophysical processes. Drs. Y. Kajiwara, T. Nakano, and K. Fujimoto read the manuscript and gave me useful critical comments. I very much appreciate Ms. M. Aizawa, Ms. N. Katayama, and Ms. K. Suga for their skillful and patient word processing. Ms. M. Shimizu and Ms. M. Komatsu of The University of Tokyo Press and Mr. Ken Kimlicka of Springer Japan edited with care the manuscripts for the Japanese version of the books *Introduction to Earth System Science* and *Introduction to Earth and Planetary System Science* and the English version of *Introduction to Earth and Planetary System Science*, respectively.

I want to dedicate this book to my wife, Midori Shikazono, and two daughters, Chikako and Hisako Shikazono, and to my parents, Naoharu and Yoshiko Shikazono, who have patiently provided understanding and moral support during the more than 30 years of my academic research and teaching.

Tokyo, Japan   Naotatsu Shikazono

Contents

Chapter 1
Introduction to Earth and Planetary System Science: A New View of the Earth, Planets, and Humans

Earth and planetary system science has been developing very rapidly recently due to (1) the establishment of plate tectonics and development of plume tectonics that provide explanations for the movement of the solid earth, (2) development of observation technologies and analysis of the earth and other planets, and (3) considerable research into the interactions between nature and humans such as the earth's environmental, resources, and disaster problems.

Plate tectonics can explain geological processes in the earth's solid surface environment, but cannot describe what occurs deep in the earth's interior, in the mantle and core. Recently, the concept of plume tectonics was proposed, resulting in a better understanding of the dynamics of the earth's interior and the genesis of earth-type planets. The development of observation and analysis technologies is accelerating, and this development may become very rapid in the near future. For example, simulations of global material circulation between the fluid and solid parts of the earth could be used to elucidate the origin and evolution of the earth and planets. The interaction between humans and nature poses more difficulties than the other two developments noted above because current methodologies do not yet offer solutions to its problems. With this purpose in mind, scientific understanding of the earth and other planets is the first step. Earth itself is not independent from any external system. It interacts with other bodies and is open to energy and mass exchange with the outside. The earth and planets did not form in isolation. The earth's history is intimately related to other planets in the solar system, and we need to know that history as well as the earth's present-day conditions to truly understand our own planet. Knowledge about the other planets has been expanding rapidly in recent years. Therefore, in this book the features of the earth, together with those of the other planets, will be described and the relationship between them will be considered. We present basic information about the earth and planets (e.g. their constituents and chemical and isotopic compositions) and describe the interactions in the earth/planets system. Recently, the earth itself as a system, particularly at the

N. Shikazono, *Introduction to Earth and Planetary System Science*,
DOI 10.1007/978-4-431-54058-8_1, © Springer 2012

surface, environment has been modified by human activity. Thus, nature–human interactions will also be considered. Then, a new view of the earth, planets and humans will be presented.

Keywords Catastrophism • Earth and planetary system science • Earth system • Earth's environment • Nature–humans interactions • Uniformitarianism

1.1   What Is the Earth and Planetary System?

1.1.1   Hierarchy in the Earth and Planetary System

The atmosphere, the hydrosphere and the geosphere (lithosphere), which are the main components of the earth, interact with each other. Thus, we can regard the earth as a system. The earth system is composed of subsystems (reservoirs) such as the atmosphere (thermosphere, mesosphere, and troposphere), the hydrosphere (seawater and terrestrial waters), and lithosphere (geosphere) (crust, mantle, and core) (Fig. 1.1). Each subsystem is divided into several parts. For example, the crust is divided into oceanic crust and continental crust. Oceanic crust is composed of basalt and ultramafic rocks, while continental crust contains granitic rocks, volcanic rocks (basalt, andesite, dacite, and rhyolite), sedimentary rocks, and metamorphic rocks. These rocks, in turn, consist of minerals, and minerals are composed of elements. Therefore, the earth system can be characterized as a hierarchy (Fig. 1.2). Most of the planets in the solar system have an atmosphere and a solid portion made of silicates and/or metals (Fig. 1.3). Water in the solar system is mostly in its solid state (ice) except in the surface environment of the earth where liquid water is present.

1.1.2   Interactions Between Subsystems

Heat and mass transfer occur between subsystems. For example, in recent years, anthropogenic emission of CO_2 gas to the atmosphere due to human activity (i.e. burning of fossil fuel) has been increasing. The increase in atmospheric CO_2 concentration causes an increase in the CO_2 concentration of seawater. This kind of mass transfer between the atmosphere and seawater is always occurring.

Heat also transfers between subsystems. For example, magma ascends from deeper parts of the earth associated with heat and mass transfer. Magma is generally generated in the upper and lower mantle. For example, generation of hot spot magma as seen in Hawaii is related to plume activity. Lower mantle exchanges heat and mass with the core. Hydrothermal solutions also transfer significant amounts of heat and mass in relatively shallower parts of the crust (generally less than 10 km). These examples illustrate how heat and mass are exchanged continuously between subsystems in the earth system.

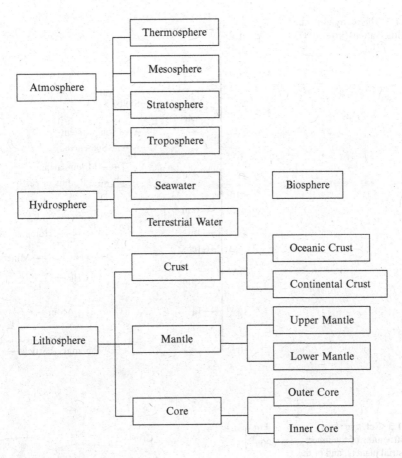

Fig. 1.1 Constituents of earth system

1.1.3   Earth as a System

It is essential to understand the general principle of a system to understand the earth and planetary system. Generally, "open" and "closed" systems are defined as those that exchange heat and mass with other systems and systems that do not, respectively (Fig. 1.4). The forces that drive heat and mass transfer include chemical potential and temperature and pressure gradients. They cause diffusion, reaction and advection (flow). The characteristics of the earth as a system and its subsystems (its compositional and spatial distributions) change irreversibly over time due to heat and mass exchange. These temporal changes are represented by differential equations. The most important purpose of earth and planetary system sciences is to elucidate the irreversible changes that the system undergoes. The first step in this

Fig. 1.2 Hierarchy of nature
(modified after Hirose 1987)

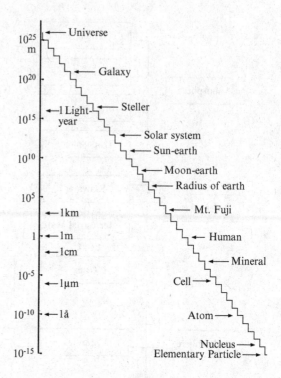

Fig. 1.3 Relative sizes
and structures of the inner,
terrestrial planets, and of the
moon (Meadow and Solomon
1984; Ernst 2000)

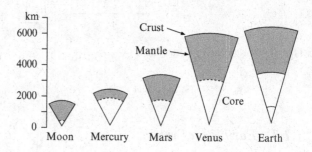

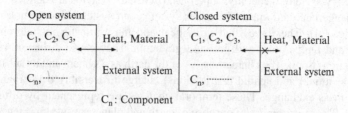

Fig. 1.4 Open system and closed system $C_1, C_2, ..., C_n, ...$: component

endeavor is to describe the natural system from various points of view (geology, chemistry, physics, biology, etc.). Then modeling of the system is required. Various models such as time-independent and dependent models, linear and nonlinear models, and deterministic and probabilistic models have been used to simulate the earth as a system. Therefore, first, we have to decide which model we should choose.

In the field of earth and planetary system sciences, the following should be investigated: (1) Different kinds of subsystems (e.g. the atmosphere, hydrosphere, and geosphere), (2) subsystem components (rocks, minerals, ions, complexes, colloids, etc.), (3) the chemical and physical states of the components of the subsystems (e.g. temperature, pressure, chemical, and isotopic composition), (4) interactions between subsystems (input and output fluxes of mass and energy), (5) the changes in subsystems' characteristics over time, and (6) the mechanisms of mass and heat transfer that determine flux caused by diffusion, reaction, and advection.

There are various interactions between subsystems. We discuss the chemical reactions between aqueous solutions and the solid phase (water–rock interactions) in detail in this text. More detailed discussions on mass transfer mechanisms can be found in "Chemistry of Earth Systems" and "Environmental Geochemistry of the Earth System" published by the author (Shikazono 1997, 2010).

1.2   What Is the Earth's Environment?

The time scale used to discuss "earth's environmental problems" is generally short compared with the history of the earth from its origin to its present state. "Earth's environment" is influenced by direct interaction between nature and humans. We call this viewpoint "earth's environment in a narrow sense", and will discuss it in Chap. 4. We will also consider the long-term, global scale earth environment, which we term the "earth's environment in a broad sense" (Fig. 1.5) in more detail.

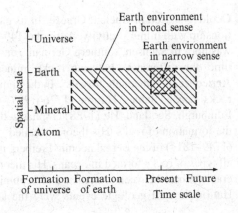

Fig. 1.5 Earth environments in a narrow sense and a broad sense in spatial scale and temporal scale diagram (Shikazono 1992)

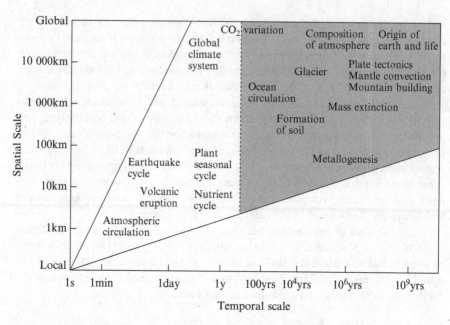

Fig. 1.6 Relationship between spatial scale and temporal scale for the geologic events and environmental problems (modified after NASA 1986)

Figure 1.6 shows the relationship between geologic events and environmental problems and the temporal and spatial scales of the earth's environment as discussed in this book.

1.3   Development of Earth and Planetary System Science

1.3.1   Geology, Geochemistry, and Geophysics

Geology started in Ancient Greece. In its early stages, geology developed in relationship with mining activity. Agricola (1494–1555), a pioneer in mineralogy, who was born in Sachsen, southern Germany, described and classified minerals. At that time, the differences between rocks, minerals and fossils was not recognized. Agricola called them all "fossils". In the eighteenth century, the geological study of rocks began. James Hutton (1726–1797) studied mineralogy and geology in Edinburgh, Scotland. He (1788) emphasized the importance of igneous activity in the formation of rocks. His theory is called "plutonism". In contrast, A. G. Werner (1749–1817) of the Bergakademie Freiberg mining school in Germany thought that all types of rocks formed in oceans. His theory is called "neptunism". After considerable debate between plutonists and neptunists, neptunism was defeated ca. 1820. Hutton thought geologic events were the accumulation of present-day geologic phenomena that change very slowly. His manner of thinking led to Charles Lyell's (1830) "uniformitarianism". Lyell's theory can be represented by the phrase "the present is a key to the past" which means that present and past geologic events are both governed by the same rules. Uniformitarianism significantly influenced Darwin's theory of evolution. "Catastrophism" is a theory from the early nineteenth century by G. D. Cuvie (1769–1832) that contrasts with uniformitarianism. An example of catastrophism is the story of Atlantis which tells us that Atlantis, which lay somewhere in the mid-Atlantic ocean sank suddenly to the seafloor one night 12,000 years ago because of an earthquake and volcanic eruption. This theory was also consistent with Noah's Ark of the Old Testament. Later, when a method for determining the earth's age was developed, it became clear that earth has been changing for around 4.6 billion years, so uniformitarianism appeared to be more plausible than catastrophism. However, recent findings of an iridium-rich layer at the unconformity between the Cretaceous and Tertiary ages and evidence of a mass extinction at that time (Alvarez et al. 1980), neither of which are known to be occurring at the present time, can be explained by catastrophism (a bolide impact) and not by uniformitarianism. After the middle of the nineteenth century, geology and mineralogy developed a great deal through the use of analytical equipment. For example, minerals came to be identified using the optical properties they exhibit under the microscope. Also, the X-ray diffraction method came to be used for determining crystal structures. Numerous chemical analyses of rocks and minerals were performed. For example, V.M. Goldschmidt (1888–1947) investigated metamorphic rocks and classified the elements they contain as siderophile (Fe, Co, Ni, etc.), chalcophile (Cu, Ag, Zn, Cd, Hg, etc.), lithophile (alkali earth, rare earth, etc.), and atmophile (H, N, C, (O), and inert gases (Ar, Xe, etc.)). Geochemistry began in the nineteenth century and developed in the early twentieth century. Harold Urey (1893–1981) established isotope geochemistry in middle of the twentieth century.

Geophysics began late in the nineteenth century. One early development in that field was when Lord Kelvin (William Thompson) (1867–1934) estimated the age of earth to be 2×10^7 years based on calculations of the thermal history starting with the earth as a molten body. In 1896, Antoine Henri Becquerel (1852–1908) discovered radioactivity. After the discovery of radium by Pierre Curie (1859–1906) and Marie Curie (1867–1934) in 1898, many radioactive elements were discovered and ages of the earth's solid materials were determined using various radioactive methods such as U-Pb dating. Since that time, considerable numbers of geophysical studies on the thermal structure of the earth and seismic and electromagnetic observations have been carried out. In the 1940s and 1950s, global weather forecasting began. Plate tectonic theory, which synthesized the solid earth sciences (geology, geophysics, etc.), was proposed in the 1960s. By the 1980s, it was an established theory that aided understanding of the dynamic solid earth (e.g. mountain building and earthquakes). Plate tectonics, however, focuses on earth's solid surface, mainly the crust, and cannot explain activity deeper in the earth's interior, in the mantle and core. Recently, the theory of plume tectonics was proposed to explain the earth's interior dynamics. It is noteworthy that igneous activities in the terrestrial planets in the solar system like Venus and Mars can be explained by this theory.

1.3.2   A New View of the Earth, Planets, and Humans: Earth and Planetary System Science

Recently, the field called "earth system science", which integrates various fields of earth science including geophysics, geology, geochemistry, geography, meteorology, oceanography, etc. has been developed. The concept of earth system science was proposed by NASA (1986). The author published "An Introduction to Earth System Science" (Shikazono 1992) and "The Chemistry of the Earth System" (Shikazono 1997, 2010). Earth system science as treated by the author is a wider field than NASA's; we treat not only earth's surface environment, which is NASA's main focus, but also earth's interior, how the earth has changed over time (history, evolution) and nature—human interactions such as use of resources and environmental problems. In the United States, starting in the late 1990s after NASA's proposal, several books on earth system sciences were published (Gradel and Crutzen 1993; Stanley 1998; Kump et al. 1999; Skinner et al. 1999; Jacobsen et al. 2000; Ernst 2000; Rollinson 2003). These books focus mainly on the global material cycle, biogeochemical cycle, ocean system and interactions between the atmosphere, biosphere, ocean, and the earth's surface environment. Earth system science as envisioned by NASA (1986) focuses on technological management of the global system. This has resulted in an insufficient understanding of humans and the biosphere and diversity in the earth system (Sacks 2002). Along with its focus on the development of technology, NASA (1986) bases its concept of the earth system on the delineation of the lithosphere, hydrosphere, and atmosphere in the earth system, Suess's (1875) idea of the biosphere coined by Vernadsky (1926, 1997), and Hutton's (1788) characterization of the earth as a "superorganism". It is important to review these early studies for their views and concepts of the earth and its subsystems.

In Japan several important works were published before NASA (1986) that regarded earth as a system. These works include Shimazu (1967, 1969), Takeuchi and Shimazu (1969), Makino (1983), Hamada (1986), and Hanya and Akiyama (1989). For example, Shimazu (1967) emphasized that nature is seamless, meaning natural phenomena interrelate with earth other. Takeuchi and Shimazu (1969) divided earth science development into three generations: first—description and classification, second—analysis and reduction, and third—integration. Third generation earth science is systems engineering on earth. In 2005, the earth and planetary union was created by the Japan Geoscience Union to establish earth and planetary system science.

The content of this book is consistent with the development of earth and planetary system science. In Chap. 2, the material aspects of the constituents (subsystems) of the earth system are described. In Chap. 3, the dynamical aspects of the solid and fluid portions of the earth (global cycles of solids and fluids, plate tectonics and plume tectonics) are considered. Chapter 4 covers basic nature–human interactions (disasters, resources, and environmental problems). In Chaps. 5 and 6, historical changes in the solar system, planets and the earth and the evolution of the earth and planetary system are reviewed and described. In Chap. 7, modern views on nature, the earth and planets are summarized, and the author's new idea of "earth's environmental co-oriented society" will be presented.

1.4 Chapter Summary

1. The entire earth can be regarded as a system that we call the "earth system". The earth system consists of several subsystems: the atmosphere, hydrosphere, biosphere, humans, and lithosphere (geosphere). The earth is open to other planets with regard to mass and energy flow so the system that comprises the earth and planets in the solar system is called the "earth and planetary system".
2. These subsystems interact with each other, and each subsystem is open to mass and energy flow to and from other systems.
3. The characteristics of the subsystems and the earth and planetary system change irreversibly over various temporal and spatial scales.
4. There are two ways of viewing the environment: considering it in a narrow sense or a broad sense. We focus here on the environment of the earth in a broad sense from its origin through the present to the future and from local to global scales.
5. Descriptive, analytical and static studies on the earth and planets were done in the early stages of earth sciences. Recently, research in the field has developed into genetic, historical, dynamic, and integrated studies. In the 1990s, earth system science was established by unifying many fields of earth sciences, and in the early 2000s, earth system science is evolving into earth and planetary system science, focusing on the interaction between humans and nature as well as the interactions between non-human subsystems and the irreversible evolution of the earth and other planets.

References

Alvarez LW, Alvarez W, Asaro F, Michel HV (1980) Extraterrestrial cause for the cretaceous-tertiary extinction. Science 208:1095–1108

Ernst WG (ed) (2000) Earth system. Cambridge University Press, Cambridge

Gradel TE, Crutzen DJ (1993) Atmospheric change—an Earth system perspective. W. H. Freeman & Co, New York

Hamada T (1986) Invitation to Earth science. University of Tokyo Press, Tokyo (in Japanese)

Hanya T, Akiyama N (1986) Human, society and Earth. Kagakudojin, Kyoto (in Japanese)

Hirose M (1987) Onion structure of nature. Kyoritsu Press, Tokyo (in Japanese)

Hutton J (1788) Theory of the Earth; or an investigation of the laws observable in the composition, dissolution and restoration of land upon the globe. R Soc Edin Trans 1:209–304

Jacobsen MC, Charlsn RI, Rodhe H, Orians GH (2000) Earth system science, International geophysics series. Elsevier Academic Press, Amsterdam, 72 pp

Kump LR, Kasting JR, Crane RG (1999) The Earth system. Rearsen Prentice Hall, Englewood Cliffs

Lyell C (1830) Principles of geology, vol 1 (London; J. Murrray, 1830: reprint, University of Chicago Press, Chicago: 1990)

Makino T (1983) Introduction to Earth—human system science. Central Press (in Japanese)

NASA (1986) Earth system science—overview—a program for global change. Earth System Sciences Committee, NASA Advisory Council, Washington

Rollinson H (2003) Early Earth systems. Blackwell, Oxford

Sacks W (2002) Planet dialectics, explorations in environment and development. Shin–Hyoron (trans: Kawamura K, Murai YI) (in Japanese)

Shikazono N (1992) Introduction to Earth system science. University of Tokyo Press, Tokyo (in Japanese)

Shikazono N (1997) Chemistry of Earth system. University of Tokyo Press, Tokyo (in Japanese)

Shikazono N (2010) Environmental geochemistry of Earth system. University of Tokyo Press, Tokyo (in Japanese)

Shimazu Y (1967) Evolution of Earth. Iwanami Shoten, Tokyo (in Japanese)

Skinner BJ, Porter SG, Botkin DB (1999) The Blue Planet. An introduction to Earth system science, 2nd edn. Wiley, New York

Stanley SM (1998) Earth system history. W. H. Freeman & Co, New York

Suess E (1875) Die Enstehung der Alpen (The Origin of the Alps). W. Braunmuller, Vienna

Takeuchi H, Shimazu Y (1969) Modern Earth science. Chikuma Shobo, Tokyo (in Japanese)

Vernadsky VI (1926) La Geochimie. Librairie. Felix Alcan

Vernadsky VI (1997) The biosphere (Langmuir DB, trans: revised and annotated by MAS McMenamin). Copernicus Books, New York

Chapter 2
Components of the Earth System

The earth and planets in the solar system are characterized by their vertical zonal structure. The structure and composition of the earth are briefly summarized below.

The earth can be divided into two parts: the fluid earth and the solid earth. The fluid earth consists of the atmosphere and hydrosphere. The atmosphere is mostly in a gaseous state, mainly consisting of N_2 and O_2. The hydrosphere is mainly water (H_2O) with many components dissolved in it. Most of the crust, mantle, and core are in a solid state. However, liquid (magma) exists in the crust, mantle, and outer core. It is primarily liquid Fe and Ni mixed with less than 10% light elements (O, H, S, Si, and K) by weight.

This chapter describes the characteristics of the fluid earth (the atmosphere and hydrosphere), the solid earth (the crust, mantle, and core), and earth's surface environment (the soils and biosphere).

The origins and formation processes of these structures are considered in Chap. 6.

Keywords Atmosphere • Biosphere • Cosmic abundant of elements • Geosphere • Ground water • Hydrosphere • Meteorites • Riverwater • Rocks • Seawater

2.1 Fluid Earth

2.1.1 The Atmosphere

The atmosphere is a thin envelope of gases surrounding the solid earth (Fig. 2.1). The temperature and composition of the atmosphere are distributed heterogeneously, and their variations can be used to subdivide the atmosphere into the troposphere, stratosphere, mesosphere, and thermosphere (Fig. 2.2). The troposphere, which extends from the surface 10–15 km upward, is a region of intense convective mixing. In the troposphere, the temperature decreases with altitude due to expansion of air heated at the surface. In the stratosphere the temperature increases with increasing

N. Shikazono, *Introduction to Earth and Planetary System Science*,
DOI 10.1007/978-4-431-54058-8_2, © Springer 2012

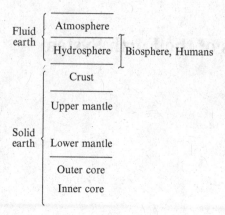

Fig. 2.1 Vertical zonal structure of earth

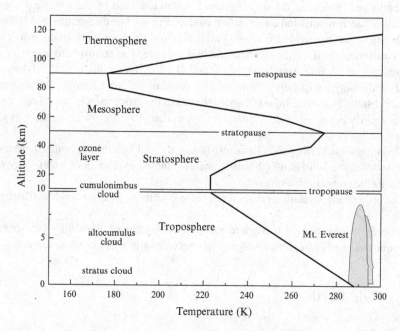

Fig. 2.2 The average vertical variation of temperature as a function of altitude in the earth's atmosphere (Chameides and Perdue 1997)

altitude up to approximately −2°C at 50 km above the surface from −50°C at 10–15 km above the surface. This is due to solar ultraviolet radiation and absorption of infrared radiation by ozone in the stratosphere. In the mesosphere, which ranges from about 50–90 km above the surface, the temperature decreases rapidly with altitude, caused by a rapidly decreasing ozone concentration.

In the thermosphere, above about 90 km from the surface, the temperature of the constituent gases again increases with altitude, reaching to 1,200°C at 500 km from

Table 2.1 Compositions of earth's atmosphere (Kawamura and Iwaki 1988)

Constant component		Variable component		
Volume (%)	Volume (ppm)	Source		Atmosphere
N_2 78.084 ± 0.004		O_3	Ultraviolet ray	0–0.07 ppm (summer)
O_2 20.746 ± 0.002				0–0.02 ppm (winter)
CO_2 0.033 ± 0.001		SO_2	Industry	0–1 ppm
Ar 0.934 ± 0.001		NO_2	Industry	0–0.02 ppm
Ne	18.18 ± 0.04	Oxidation of biogenic CH_2O		Not certain
He	5.24 ± 0.004			
Kr	1.14 ± 0.014	I_2	Industry	Less than 10^{-4} g/m^3
Xe	0.087 ± 0.001	NaCl	Seasalt	10^{-4} g/m^3
H_2	0.5	NH_3	Industry	0-trace
CH_4	2	CO	Industry	0-trace
N_2O	0.5 ± 0.1	H_2O	Evaporation	0.35 g/m^3

the surface. At that height, solar ultraviolet radiation photons associated with light of wavelength less than about 120 mm are largely absorbed by O_2 and N_2 in the thermosphere and converted to heat.

The atmosphere is mostly composed of three elements—nitrogen (N), oxygen (O), and argon (Ar)—with minor concentrations of other elements and compounds (Table 2.1).

The concentration of the major components (N_2, O_2, and Ar) is relatively constant, mixed by convection currents from the surface to an altitude of about 80 km. The concentration is particularly uniform up to about 12 km above the surface. In contrast, at 100 km above the surface, the concentration of N_2 is smaller, while H_2 and He concentrations are larger than at lower altitudes.

Data available on the concentrations of minor and trace gases—H_2, N_2O, Xe, CO, O_3, NH_3, CH_2O, NO, NO_2, SO_2, chlorofluorocarbons ($CFCl_3$, CF_2Cl_2), carbon tetrachloride (CCl_4), and methyl chloride (CH_3Cl)—in the atmosphere is very sparse compared with information about the major components. Recently, data on minor and trace components consisting of aerosols and solid and liquid particles in suspension in the air have been obtained. Aerosols include particles derived naturally from weathering, volcanic ash, marine salt (NaCl), etc., sulfate particles from dimethyl sulfide (DMS) and anthropogenic particles from the burning of biomass and fossil fuels. Much of the minor and trace components and aerosols are of human origin, such as emissions from factories, cars, power plants, etc. They are heterogeneously distributed in the atmosphere and so are called variable components. The concentrations of the major components are relatively constant so they are called constant components. The average concentrations of the major and minor components in the atmosphere are shown in Table 2.1. The heterogeneous concentrations of the minor components are caused by their varying sources and high rates of dissolution into rainwater.

For example, SO_2 gas contained in emissions from factories is soluble in rainwater, as is HCl. These gases are removed from the atmosphere quickly. The solubility of

Table 2.2 Abundance of water in the hydrosphere (modified after Skinner 1976)

Place	Amount (l)	(%)
Fresh lake	125×10^{15}	0.009
Saline lake and island sea	104×10^{15}	0.008
Average riverwater	1×10^{15}	0.0001
Soil water	67×10^{15}	0.005
Shallow ground water (less than 80 m depth)	$4,170 \times 10^{15}$	0.31
Deep ground water	$4,170 \times 10^{15}$	0.31
Glacier	$29,000 \times 10^{15}$	2.15
Atmosphere	13×10^{15}	0.001
Ocean	$1,320,000 \times 10^{15}$	97.2

Table 2.3 Average composition of major elements in seawater (salinity 35‰) (Nishimura 1991)

Cl^-	19.353
Na^+	10.766
SO_4^{2-}	2.708
Mg^{2+}	1.293
Ca^{2+}	0.413
K^+	0.403
CO_3^{2-}	0.142
Br^-	0.674

[a]HCO_3^- is included in CO_3^{2-}

gases in water is highly dependant on the gas species. Solubility differences also cause minor and trace components in the atmosphere to be heterogeneously distributed.

2.1.2   The Hydrosphere

The hydrosphere consists of all the water on the earth. Earth's water is mostly composed of seawater, terrestrial waters, including riverwater, lakewater, ground water, etc., and frozen water in the polar ice caps, ice sheets, and glaciers. The earth's water is 97% is seawater and 2% frozen water (Table 2.2).

2.1.2.1   Seawater

Seawater has a relatively constant composition (Table 2.3) compared with other geologic bodies (e.g. rocks, terrestrial waters). Seawater near the coasts and in the open sea have different compositions. The salinity of seawater near the coast is lower than that in the open sea mainly because of riverine input.

Seawater in closed areas like bays or inland seas is generally anoxic due to the consumption of O_2 dissolved in the seawater by the oxidation of organic matter.

The temperature of seawater varies as a function of depth. In shallower zones (from 100 to 1,000 m deep) the water temperature is high, while in deeper zones it is cold. The temperature in shallow seawater changes rapidly between 100 and 1,000 m in depth. The shallow layer above the zone where the temperature change occurs is called surface layer seawater and the layer below it is called deep layer seawater.

The average composition of the major elements in seawater is presented in Table 2.3. Na^+ is the most abundant cation but Ca^{2+}, Mg^{2+}, and K^+ are also found in high concentrations. Cl^- is the most highly concentrated anion. The concentration of sulfate (SO_4^{2-}) is also high. The sum of the concentrations of the six components listed above exceeds 99.81% of the total concentration of dissolved species. Seawater concentrations are relatively constant compared with the concentrations in terrestrial waters, but this varies by location and depth. For example, the pH of surface seawater is about 8.2, but the pH decreases with depth, because of the formation of calcium carbonate by marine organisms like coral, foraminifers, and urchins, which proceeds according to the following reactions:

$$Ca^{2+} + 2HCO_3^- \rightarrow CaCO_3 + H_2O + CO_2$$

$$CO_2 + H_2O \rightarrow H^+ + HCO_3^-$$

The dissolution of calcium carbonate, which is written as

$$CaCO_3 + H^+ \rightarrow Ca^{2+} + HCO_3^-$$

causes the pH to increase.

The concentrations of the minor elements vary with depth. There are three types of this variation. The first is where a concentration is constant with depth (e.g. U and Mo). This type of elements forms stable complexes. The second variation is where a concentration increases with depth. Most elements (30 elements) belong to this type. Their chemical features are similar to nutrient elements, (N, P, and Si), which are used in biological processes, so this is called the nutrient type. The third variation is represented by Al and Pb, which are abundant in the shallow layer and depleted in the deep layer. They are removed by scavenging and their residence time (the ratio of the total quantity of any component to its rate of input or output) is short (Al: 10^2 years, Pb: $10^{2.6}$ years) (Holland 1978). The concentrations of elements in seawater are governed mainly by biological activity, interaction with sediments, and the reactions involving CO_2.

2.1.2.2 Riverwater

Ca^{2+} has the highest concentration among all the cations in riverwater (Table 2.4). Ca^{2+} is derived from the dissolution of calcium carbonate (calcite) in limestone and Ca silicates in rocks. This is represented by the reaction

$$(CaO) + 2H_2CO_3 \rightarrow Ca^{2+} + 2HCO_3^- + H_2O$$

where (CaO) is the CaO component in minerals.

Table 2.4 Average chemical composition of riverwater (mg/l). Japanese riverwater; Kobayashi (1960), World riverwater; Livingstone (1983)

	Japan	World		Japan	World
Na^+	6.7	6.3	HCO_3^-	31.0	58.4
K^+	1.19	2.3	SO_4^{2-}	10.6	11.2
Mg^{2+}	1.9	4.0	Fe	0.24	0.67
Ca^{2+}	8.8	15.0	SiO_2	19.0	13.1
Cl^-	5.8	7.8			

Mg^{2+}, K^+, and Na^+ are also derived mainly from silicates. HCO_3^- is the most abundant anion. Two thirds of the HCO_3^- in riverwater is derived from atmospheric CO_2, plant decomposition, and photosynthesis, and a small proportion comes from the oxidation of organic matter in sediments.

Riverwater originates as rainwater, ground water, spring water, and surface water. The mixing ratio of these waters determines the chemical composition of riverwater. The chemical composition of major elements such as Si in ground water and spring water are considerably influenced by water–rock interaction and the geology of the watershed. Rainwater penetrates deeper underground, which brings it to react with the surrounding soils and rocks. During these reactions, cations like Ca^{2+} and Na^+ are released from the soils and rocks. The concentrations of the cations are largely dependant on the kinds of rocks and minerals with which the water comes in contact. In contrast, if the time that it takes for rainwater to reach the river is very short, the chemical composition of the riverwater is not so different from that of the rainwater. In watershed where limestone is widely distributed, the chemical composition of riverwater comes close to an equilibrium state with respect to carbonates.

Seawater salt particles in the atmosphere influence the chemical composition of riverwater near the sea coast. For example, the concentrations of Na^+ and Cl^- in the riverwater on Japanese Islands surrounded by the ocean are high (Table 2.4). Hot springs and volcanic gases related to volcanic activity also influence riverwater chemistry. For example, the acidity of riverwater is enhanced by inputs from hot springs, volcanic gases, and sublimates such as native sulfur. Anthropogenic activities influence the chemical composition of riverwater in urban area to a considerable degree due to inputs of various waste waters and wastes. Acid rain also has an effect, reducing the pH of riverwater (Sect. 4.4.3).

2.1.2.3 Ground Water

Ground water is water that originated as rainwater that penetrated underground. During the penetration and flow of ground water, water reacts with the surrounding rocks and soils, causing changes in the chemical composition of the water (Fig. 2.3). The chemical composition of ground water is determined by the kinds of rocks and minerals it encounters, reaction time, and flow rate. The reaction time and flow rate depend on chemical and physical properties of the rocks, especially grain size,

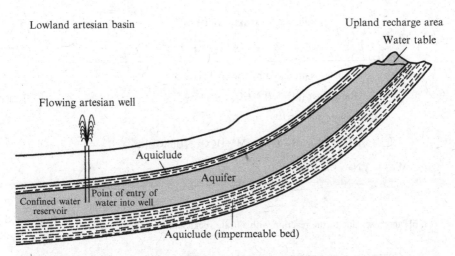

Lowland artesian basin

Upland recharge area

Water table

Flowing artesian well

Aquiclude

Aquifer

Confined water reservoir

Point of entry of water into well

Aquiclude (impermeable bed)

Fig. 2.3 A confined water reservoir is created where water enters an aquifer sandwiched between two confining aquifers. The artesian well flows in response to the pressure difference between the height of the water table in the recharge area and the bottom of the well before the well as drilled (Holland and Petersen 1995)

porosity, and permeability. These properties vary widely so the chemical composition of ground water are different in different areas. The chemical properties of rainwater also control the chemical composition of ground water. Rainwater is usually slightly acidic (pH ~ 5.6) due to the dissolution of atmospheric CO_2. Recently, acid rain caused by human activities such as the burning of fossil fuels has become common (Sect. 4.4.3). This acid rain penetrates underground, resulting in the release of various elements from minerals and enhancement of pH by the reaction (Fig. 2.4)

$$MO + 2H^+ \rightarrow M^{2+} + H_2O$$

where MO is the MO component in a mineral, and M is a divalent cation (e.g. Ca^{2+} and Mg^{2+}). and

$$CaCO_3 + H_2CO_3 \rightarrow Ca^{2+} + 2HCO_3^-$$

$$MgCO_3 + H_2CO_3 \rightarrow Mg^{2+} + 2HCO_3^-$$

Deep ground water is sometimes in equilibrium with rocks and minerals, whether they be silicates (montmorillonite, kaolinite, feldspar, and opal) or carbonates (calcite and dolomite). Ground water with slow flow rates and long reaction times tends to achieve equilibrium, while the high flow rates and short reaction times in shallow ground water systems do not.

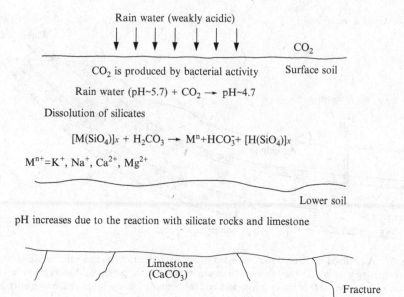

Rain water (weakly acidic)

CO_2

CO_2 is produced by bacterial activity   Surface soil

Rain water (pH~5.7) + CO_2 → pH~4.7

Dissolution of silicates

$$[M(SiO_4)]x + H_2CO_3 → M^n + HCO_3^- + [H(SiO_4)]x$$

$M^{n+} = K^+, Na^+, Ca^{2+}, Mg^{2+}$

Lower soil

pH increases due to the reaction with silicate rocks and limestone

Limestone
($CaCO_3$)

Fracture

$$CaCO_3 + H_2CO_3 → Ca^{2+} + 2HCO_3^-$$
$$MgCO_3 + H_2CO_3 → Mg^{2+} + 2HCO_3^-$$

Fig. 2.4 Penetration of rainwater into soil and rocks and reactions between water and rocks

2.1.2.4 Classification of Earth's Water

Various kinds of water are distributed in the earth system. They include seawater, riverwater, ground water, lakewater, soilwater, water in organisms, water in minerals, and rainwater. Water exists not only in the hydrosphere; but also in the atmosphere as rainwater, snow, fog, hail, and water vapor; the lithosphere as water adsorbed onto minerals, water in crystal structures, and fluid inclusions in crystals; and the biosphere.

The abundance of types of water in earth's system is shown in Table 2.2. The most abundant is seawater (97%). The other relatively abundant types are terrestrial water (ground water, riverwater, and lakewater) and ice (glaciers and ice sheets). The other types occur in very small amounts, but they are used in various fields as water resources (Sect. 4.4.5).

2.2 Solid Earth (the Geosphere)

2.2.1 The Earth's Interior Structure

Direct and indirect approaches can be used to estimate the composition and structure of the earth's interior. Drillholes, xenoliths in volcanic rocks, and geological surveys of the earth's surface tell us directly about the composition of the earth's interior.

The deepest people have drilled is about 10 km. However, there is a plan to drill deeper for the purpose of seismic study and exploration of mineral and fossil fuel resources in Japan. If this plan is implemented, we may get more information on deeper parts of the earth. However, drilling data cannot give us three-dimensional information on the earth's interior. A xenolith is a rock fragment contained in igneous rocks that are derived from magma deeper in the interior. We can know about mantle material from xenoliths. They are usually composed of olivine, and so are called olivine nodules. The distribution of olivine nodules is very restricted, and so they do not provide information on the composition of the entire mantle.

In contrast, indirect approaches give us useful data on large parts of the earth's interior. These include high temperature and pressure experiments, and gravity, seismic, and electric conductivity measurements.

A phase diagram for minerals can be constructed based on high temperature and pressure experiments examining the stability of minerals. From the phase diagrams and the mineralogy of xenoliths, we can estimate the temperature and pressure conditions under which the xenolith material formed. Igneous rocks form by the solidification of magma. Data from the combination of high temperature and pressure experiments with mineralogy can provide the information necessary to deduce the temperature and pressure of the magma from which igneous rocks formed. However, magmatic composition is not identical to that of the mantle material. It depends on the degree of melting, temperature, pressure, and source materials.

Seismic waves are the most useful tool for deciphering the structure of earth's interior. An earthquake generates various kinds of waves. These seismic waves include surface waves, which travel only across the surface, and body waves which travel through the earth's interior. Among the waves, P waves (primary waves) and S waves (secondary waves) are useful in understanding the earth's interior. The velocities of P and S waves depend on the density and elastic constants of the material through which they pass, and they are subject to reflection and refraction at surfaces of discontinuity. From the relationship between the velocities of P and S waves and depth, we know discontinuities exist in the earth's interior (Fig. 2.5). The velocities of P and S waves are represented by the equations

$$Vp^2 = (Ks + 4/3\mu)/\rho$$

$$Vs^2 = \mu/\rho$$

where Vp and Vs are the velocities of P and S waves, respectively, ρ is the density of the material, Ks is the bulk modulus, and μ is the rigidity. Using these equations, seismic data and some plausible assumptions of the parameter values, we can deduce the density and pressure distributions in the earth's interior.

2.2.2   The Composition of the Crust

The crust extends from the surface of the earth to Mohorovicic's (Moho's) discontinuity, which is the first plane of unconformity, or boundary between the mantle

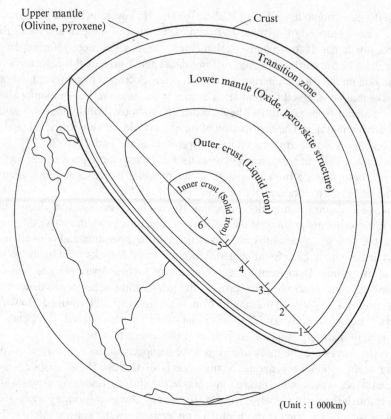

Upper mantle
(Olivine, pyroxene)

Crust

Transition zone

Lower mantle (Oxide, perovskite structure)

Outer crust (Liquid iron)

Inner crust (Solid iron)

6

5

4

3

2

1

(Unit : 1 000km)

Fig. 2.5 Earth's interior structure (modified after Sugimura et al. 1988)

and crust. The thickness of the crust varies considerably. Oceanic crust is thin, between 10 and 13 km deep. Below active mountain belts, however, it can descend to as deep as 65 km from an average of about 35 km.

The average chemical composition of upper continental crust has been estimated from large amounts of analytical data on surface rocks and large river sediments (Table 2.5). From these data it is certain that the average composition of continental crust is approximately granite:basalt = 1:1.

2.2.3   Composition of the Mantle

The mantle reaches from Mohorovicic's discontinuity to a depth of 2,900 km. It can be divided into the upper and lower mantles based on seismic velocity measurements. The boundary between the upper and lower mantle lies at a depth of about 680 km. There is a low velocity zone where the velocity of seismic waves is slower than in the other parts of mantle. This zone is considered to be partially molten. The average

Table 2.5 Average chemical composition of major elements in upper continental crust (wt%) (Rudnick and Gao 2004)

	Clarke and Washington (1924)	Goldschmidt (1933)	Condie (1993)	Taylor and McLennan (1985)	Wedepohl (1995)	Rudnick and Gao (2004)
SiO_2	60.30	62.22	67.0	65.89	66.8	66.62
TiO_2	1.07	0.83	0.56	0.50	0.54	0.64
Al_2O_3	15.65	16.63	15.14	15.17	15.05	15.40
FeO^t	6.70	6.99	4.76	4.49	4.09	5.04
MnO	0.12	0.12	-	0.07	0.07	0.10
MgO	3.56	3.47	2.45	2.20	2.30	2.48
CaO	5.18	3.23	3.64	4.19	4.24	3.59
Na_2O	3.92	2.15	3.55	3.89	3.56	3.27
K_2O	3.19	4.13	2.76	3.89	3.19	2.80
P_2O_5	0.31	0.23	0.12	0.15	0.15	0.15

FeO^t: Total iron ($FeO + Fe_2O_3$)

Table 2.6 Major element chemical composition of primitive mantle (wt%) (Palme and O'Neil 2004)

	Ringwood (1979)	McDonough and Sun (1995)	Allegre et al. (1995)	Palme and O'Neil (2004)
MgO	38.1	37.8	37.77	36.77 ± 0.44
Al_2O_3	3.3	4.4	4.09	4.49 ± 0.37
SiO_2	45.1	45.0	46.12	45.40 ± 0.30
CaO	3.1	3.5	3.23	3.65 ± 0.31
FeO^t	8.0	8.1	7.49	8.10 ± 0.05
Total	97.6	98.8	98.7	98.41 ± 0.10

Primitive mantle: mantle before the separation of crust, FeO^t: Total iron ($FeO + Fe_2O_3$)

chemical composition of the mantle has been estimated from analytical data on igneous rocks and xenoliths, high temperature and pressure experiments, and the velocity of seismic waves. Mineral phases change with depth in the mantle, varying with temperature and pressure (Fig. 2.5). For example, quartz (SiO_2) and pyroxene ($MgSiO_3$) change to stishovite (SiO_2) and perovskite ($MgSiO_3$), respectively, under higher pressures. Estimates of the average chemical composition of mantle calculated by previous researchers are in general agreement with each other (Table 2.6). It has been generally inferred that the chemical composition of the mantle is relatively uniform. However, it was found that isotopic compositions of, for example, Pb and Sr in the mantle are heterogeneous, suggesting varying compositions of major and minor elements in the mantle. Thus three-dimensional variations in chemical and isotopic compositions of the mantle and their causes need to be investigated.

2.2.4 The Composition of the Core

The core is the part of the earth deeper than about 2,890 km below the surface. The core is divided into an inner core and an outer core. S waves do not travel through the outer core, indicating that the outer core is in a liquid state. The inner core is

Table 2.7 Estimated composition of major elements (wt%) and minor elements (ppm) core in (McDonough 2004)

H	600	Zn	0	Pr	0
Li	0	Ga	0	Nd	0
Be	0	Ge	20	Sm	0
B	0	As	5	Eu	0
C (%)	0.12	Se	8	Gd	0
N	75	Br	0.7	Tb	0
O (%)	0	Rb	0	Dy	0
F	0	Sr	0	Ho	0
Na (%)	0	Y	0	Er	0
Mg (%)	0	Zr	0	Tm	0
Al (%)	0	Nb	0	Yb	0
Si (%)	6.0	Mo	5	Lu	0
P (%)	0.20	Ru	4	Hf	0
S (%)	1.90	Rh	0.74	Ta	0
Cl	200	Pd	3.1	W	0.47
K	0	Ag	0.15	Re	0.23
Ca (%)	0	Cd	0.15	Os	2.8
Sc	0	In	0	Ir	2.6
Ti	0	Sn	0.5	Pt	5.7
V	150	Sb	0.13	Au	0.5
Cr (%)	0.90	Te	0.85	Hg	0.05
Mn	300	I	0.13	Tl	0.03
Fe (%)	85.5	Cs	0.065	Pb	0.4
Co	0.25	Ba	0	Bi	0.03
Ni (%)	5.20	La	0	Th	0
Cu	125	Ce	0	U	0

solid, composed predominantly of iron and nickel (Table 2.7). The density of the inner core is 10% lower than an iron–nickel alloy, indicating that it contains about 8–10% light elements including S, H, O, and Si.

2.2.5   The Lithosphere and Asthenosphere

Seismic studies have revealed that S wave velocity decreases rapidly at a depth of about 60–80 km depth. This range is called the low velocity zone. Electric conductivity also changes here. The lithosphere and asthenosphere are above and below the low velocity zone, respectively (Fig. 2.6). The part below the asthenosphere is called the mesosphere (Fig. 2.6). The lithosphere is divided into more than ten plates and consists of the crust and upper mantle. The lithosphere varies from about 40–100 km thick, averaging about 70 km. The structure and composition of the lithosphere lying under oceanic crust is relatively constant, but continental crust is composed of

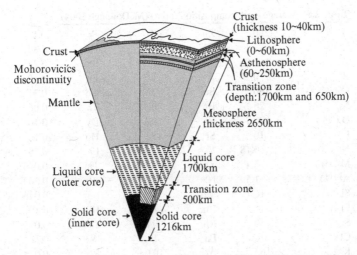

Fig. 2.6 The structure of earth's interior (modified after Sullivan 1974). *Right side*: present-day view; *left side*: previous view

various (igneous, metamorphic, and sedimentary) kinds of rocks. The classification of the lithosphere, asthenosphere, and mesosphere is consistent with the concept of plate tectonics and is different from the division of the earth's interior into only the crust, mantle, and core. Plate tectonics is described in Sect. 3.4.1.3.

2.2.6 Average Composition of the Earth as a Whole

It is generally said that the earth was formed from chondrites. Chondrites are mainly composed of silicates, iron sulfide (troilite, FeS), and iron–nickel alloys. They are primitive meteorites and are characterized by having chondrules, which are small particles of silicates. We can deduce the average composition of the earth as a whole from models that begin with the origin of earth and estimates of the relative amounts and compositions of the mantle and core (Tables 2.6 and 2.7). The alkali element contents of the earth are relatively smaller than in chondrites and some minor elements differ from chondrites. However, the earth's average elemental abundance has been deduced mainly based on chondrite models (Table 2.8).

By weight, 90% of the entire earth is made up of Fe, O, Si, and Mg. Other elements over 1% by weight include Na, Ca, Al, and S. K, Cr, Co, P, Mn, and Ti are each 0.1– 1% by weight. Thus the earth is essentially made up of these 14 elements. This is similar to the overall abundance in the cosmos and the chemical features of meteorites. Fe and Si occur in cosmic abundances and Mg, Ni, Na, K, and Al are similar to their abundance in meteorites. Volatile elements such as He, H, Ar, Cr, and Xe are very small in abundance compared with the cosmos, as are C, N, and O. The depletion of volatile elements in the earth is possibly due to intense degassing at an early stage of the formation of the earth and degassing from source materials before its formation.

Table 2.8 Average earth composition (ppm) (McDonough 2004)

H	260	Zn	40	Pr	0.17
Li	1.1	Ga	3	Nd	0.84
Be	0.05	Ge	7	Sm	0.27
B	0.2	As	1.7	Eu	0.10
C	730	Se	2.7	Gd	0.37
N	25	Br	0.3	Tb	0.067
O (%)	29.7	Rb	0.4	Dy	0.46
F	10	Sr	13	Ho	0.10
Na (%)	0.18	Y	2.9	Er	0.30
Mg (%)	15.4	Zr	7.1	Tm	0.046
Al (%)	1.59	Nb	0.44	Yb	0.30
Si (%)	16.1	Mo	1.7	Lu	0.046
P	715	Ru	1.3	Hf	0.19
S	6,350	Rh	0.24	Ta	0.025
Cl	76	Pd	1	W	0.17
K	160	Ag	0.05	Re	0.075
Ca (%)	1.71	Cd	0.08	Os	0.9
Sc	10.9	In	0.007	Ir	0.9
Ti	810	Sn	0.25	Pt	1.9
V	105	Sb	0.05	Au	0.16
Cr	4,700	Te	0.3	Hg	0.02
Mn	800	I	0.05	Tl	0.012
Fe (%)	32.0	Cs	0.035	Pb	0.23
Co	880	Ba	4.5	Bi	0.01
Ni (%)	1.82	La	0.44	Th	0.055
Cu	60	Ce	1.13	U	0.015

The crust of the earth is characterized by depleted amounts of Mg and Cr and enriched Al, K, Na, and Ca, resulting from enrichment of feldspar in the crust.

2.2.7 The Cosmic Abundance of Elements

The cosmic abundance of elements is considered to be approximately equal to the elemental abundance in the primitive solar system. The cosmic abundance is estimated from chemical compositions of Cl chondrites and stars, especially the atmosphere of the sun (Fig. 2.7). Cl chondrites are the most primitive meteorites, so their composition reflects that of the primitive solar nebula, but the chondrites are depleted in volatiles. Therefore, we estimate the cosmic abundance of volatiles based on atmospheric data from the sun. The uncertainties of these data are large, but the cosmic abundance can be estimated using Suess's nuclear systematics, in which the logarithm of the abundance of odd numbered elements varies smoothly with A (mass number) when $A \geq 50$ (Table 2.8, Fig. 2.8).

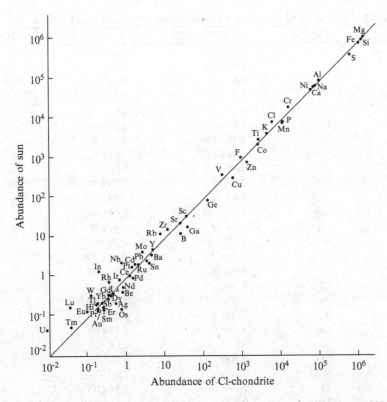

Fig. 2.7 Comparison of elemental abundance of sun and that of C1-chondrite (Ebihara 2006)

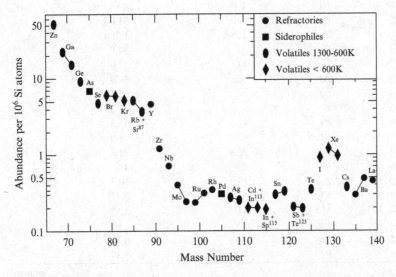

Fig. 2.8 Elemental abundance and odd number elements (Anders and Ebihara 1982). $_{14}Si = 1 \times 10^6$

Table 2.9 Cosmic abundances of the elements (atoms/10^6Si) (Anders and Ebihara 1982)

	Element	Cameron (1982)	Anders and Ebihara (1982)
1	H	2.66×10^{10}	2.72×10^{10}
2	He	1.8×10^9	2.18×10^9
3	Li	60	59.8
4	Be	1.2	0.78
5	B	9	24
6	C	1.11×10^7	1.21×10^7
7	N	2.31×10^6	2.48×10^6
8	O	1.84×10^7	2.01×10^7
9	F	780	843
10	Ne	2.6×10^6	3.76×10^6
11	Na	6.0×10^4	5.70×10^4
12	Mg	1.06×10^6	1.075×10^6
13	Al	8.5×10^4	8.49×10^4
14	Si	1.00×10^6	1.00×10^6
15	P	6,500	1.04×10^4
16	S	5.0×10^5	5.15×10^5
17	Cl	4,740	5,240
18	Ar	1.06×10^5	1.04×10^5
19	K	3,500	6,770
20	Ca	6.25×10^4	6.11×10^4
21	Sc	31	33.8
22	Ti	2,400	2,400
23	V	254	295
24	Cr	1.27×10^4	1.34×10^4
25	Mn	9,300	9,510
26	Fe	9.0×10^5	9.00×10^5
27	Co	2,200	2,250
28	Ni	4.78×10^4	49.3×10^4
29	Cu	540	514
30	Zn	1,260	1,260
31	Ga	38	37.8
32	Ge	117	118
33	As	6.2	6.79
34	Se	67	62.1
35	Br	9.2	11.8
36	Kr	41.3	45.3
37	Rb	6.1	7.09
38	Sr	22.9	23.8
39	Y	4.8	4.64
40	Zr	12	10.7
41	Nb	0.9	0.71
42	Mo	4.0	2.52
44	Ru	1.9	1.86
45	Rh	0.40	0.344
46	Pd	1.3	1.39

(continued)

Table 2.9 (continued)

	Element	Cameron (1982)	Anders and Ebihara (1982)
47	Ag	0.46	0.529
48	Cd	1.55	1.69
49	In	0.19	0.184
50	Sn	3.7	3.82
51	Sb	0.31	0.352
52	Te	6.5	4.91
53	I	1.27	0.90
54	Xe	5.84	4.35
55	Cs	0.39	0.372
56	Ba	4.8	4.36
57	La	0.37	0.448
58	Ce	1.2	1.16
59	Pr	0.18	0.174
60	Nd	0.79	0.836
62	Sm	0.24	0.261
63	Eu	0.094	0.0972
64	Gd	0.42	0.331
65	Tb	0.076	0.0589
66	Dy	0.37	0.398
67	Ho	0.092	0.0875
68	Er	0.23	0.253
69	Tm	0.035	0.0386
70	Yb	0.20	0.243
71	Lu	0.035	0.0369
72	Hf	0.17	0.176
73	Ta	0.020	0.0226
74	W	0.30	0.137
75	Re	0.051	0.0507
76	Os	0.69	0.717
77	Ir	0.72	0.660
78	Pt	1.41	1.37
79	Au	0.21	0.186
80	Hg	0.21	0.52
81	Tl	0.19	0.184
82	Pb	2.6	3.15
83	Bi	0.14	0.144
90	Th	0.045	0.0335
92	U	0.027	0.0090

Table 2.10 Classification of
meteorites (modified after
Matsuhisa and Akagi 2005)

chondrite	Numbers
Carbonaceous chondrite (C chondrite)	
C1	5
CM	171
CR	78
CO	85
CV	49
CK	73
CH	11
CB	5
Ordinary chondrite (O chondrite)	
H	6,962
L	6,213
LL	1,048
Enstatite chondrite (E chondrite)	
EH	125
EL	38
R chondrite	19
K chondrite	3
Differentiated meteorite	
A chondrite	554
Ironstone meteorite	110
Iron meteorite	770
Mercurian meteorite (SNC)	26
Moon meteorite	18

2.2.8 Meteorites

Many meteorites have been discovered on the surface of the earth. Many studies on
meteorites have revealed their composition, texture and origins. Meteorites are classified into several groups by composition and texture (Table 2.10). The main types
are (1) Chondrites (93%), (2) Irons, averaging 98% metal (6%), (3) Stony irons,
averaging 50% metal and 50% silicate (1%), and (4) Aerolites or stones.

There are three types of chondrites: carbonaceous, ordinary, and enstatite chondrite. They contain chondrules, which are small particles of silicate several millime-
ters in diameter. Chondrites are ultramafic in composition and generally consist of
olivine and pyroxene, but sometimes glass carbonaceous chondrites contain CAIs
which are calcium- and aluminum-rich and Si-poor inclusions. CAIs are thought to
have been heated to elevated temperatures, but the source of the heat required to do
so is uncertain.

The composition of the earth is similar to that of chondrites, indicating that the
earth originated mainly from materials similar to them. Normal chondrites are the
most common and can be subclassified into H, L, and LL types. Carbonaceous chondrites contain appreciable amounts of volatiles like carbon, water, and organic matter

Fig. 2.9 Structure of Si–O tetrahedron

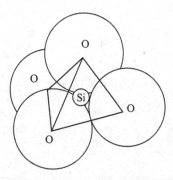

in addition to silicates. These volatiles are similar in composition to the atmosphere of the sun, indicating that it is the most primitive chondrite in the solar system.

Some chondrites do not contain chondrules, such as evolved stony chondrites (stony irons and irons). Some chondrites discovered on the earth have been determined to come from Mars and the Moon.

Iron meteorites consist of a nickel (Ni)–iron (Fe) alloy and are characterized by Widmanstätten figures that are intergrowths of lamellae of Ni-rich taenites (Ni and Fe) in Ni-poor Ni–Fe alloys (kamacite).

Stony iron meteorites are composed of Ni–Fe alloys and silicates in approximately equal amounts. The dominant silicate is olivine. Pyroxene occurs sometimes in small amounts. The silicates are distributed in the iron–nickel alloy.

2.2.9 Materials in the Solid Earth

2.2.9.1 Minerals

The solid part of the earth is composed of rocks made from minerals. A mineral is defined as a natural, inorganic, crystalline substance.

Most rocks are composed of crystalline minerals. However, some are noncrystalline and amorphous materials like volcanic glass, although these occur in small amounts.

The most abundant element in the crust and mantle is oxygen (O). Next is silicon (Si). Oxygen combines with silicon to form SiO_4^{4-}, which is the basic unit of silicate minerals. SiO_4^{4-} is tetrahedral in form. It has a pyramidal structure with four sides, composed of a central silicon ion (Si^{4+}) surrounded by four oxygen ions (Fig. 2.9).

Usually, foreign cations such as Ca^{2+}, Mg^{2+}, Na^+, and K^+ are contained in silicate crystals, together with anions such as SiO_4^{4-}.

Silicate minerals are classified into several groups according to their degree of polymerization as measured by the number of nonbridging oxygens (those bonded to just one Si^{4+}) (Table 2.1). They include the monomer silicates olivine and garnet, the chain silicates that make up the pyroxene group, double-chain silicates known

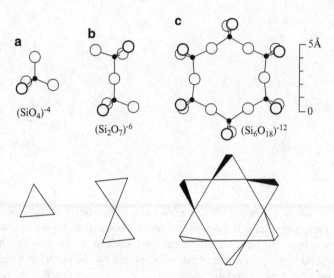

Fig. 2.10 Combination models of (**a**) independent silicon tetrahedra such as occur in olivine and garnet, (**b**) paired tetrahedra as found in epidotes, and (**c**) six-member tetrahedral rings character-istic of the mineral beryl. *Small black spheres* represent silicon, *large open spheres* represent oxygen (Ernst 1969, 2000)

Table 2.11 The structural classification of silicate minerals (Mason 1958)

Classification	Structural arrangement	Silicon:oxygen ratio	Examples
Nesosilicates	Independent tetrahedra	1:4	Forsterite, Mg_2SiO_4
Sorosilicates	Two tetrahedra sharing one oxygen	2:7	Akermanite, $Ca_2MgSi_2O_7$
Cyclosilicates	Closed rings of tetrahedra each sharing two oxygens	1:3	Benitoite, $BaTiSi_3O_9$ Beryl, $Al_2Be_3Si_6O_{18}$
Inosilicates	Continuous single chains of tetrahedra each sharing two oxygens. Continuous double chains of tetrahedra sharing alternately two and three oxygens	1:34:11	Pyroxene, e.g., enstatite, $MgSiO_3$ Amphiboles, e.g., anthophyllite, $Mg_7(Si_4O_{11})_2(OH)_2$
Phyllosilicates	Continuous sheet of tetrahedra each sharing three oxygens	2:5	Talc, $Mg_3Si_4O_{10}(OH)_2$ Phlogopite, $KMg_3(AlSi_3O_{10})(OH)_2$
Tektosilicates	Continuous framework of tetrahedra each sharing four oxygens	1:2	Quartz, SiO_2 Nepheline, $NaAlSiO_4$

as the amphibole group, sheet silicates that make up the mica group and clay miner-als, and framework silicates including quartz and the feldspar group (Fig. 2.10).

Representative silicate minerals are quartz, feldspar, pyroxene, olivine, mica, amphibole, and garnet. These minerals occur heterogeneously in earth's interior. Feldspar and quartz are the most common minerals in continental crust. In the upper mantle, pyroxene, olivine and garnet occur abundantly.

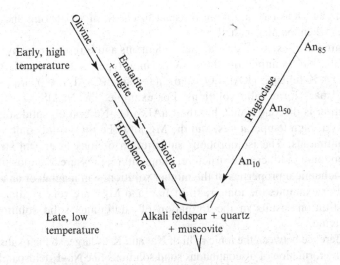

Fig. 2.11 The reaction series of mineral crystallization from a cooling, chemical differentiating magma. The discontinuous sequence of minerals is illustrated on the *left*, the continuous sequence on the *right*, An: anorthite (Bowen's reaction series) (Ernst 1969)

Fig. 2.12 Weathering sequence of silicate minerals (Goldich's weathering series) (Goldich 1938; Lasaga 1981)

Deeper in the earth's interior, denser minerals, instead of above silicates, are stable. For example, stishovite and perovskite appear in deeper regions instead of quartz and garnet, respectively. Olivine changes to perovskite. In stishovite and perovskite crystals, SiO_4^{4-} is enclosed by six O^{2-} ions (six coordinations) and the basic unit is octahedral.

The crystallization sequence of silicate minerals in magma, known as Bowen's reaction series (Fig. 2.11) and their weathering sequence, known as Goldich's weathering series (Fig. 2.12) can be explained by the crystal structure of silicate minerals. Silicate minerals with a high degree of polymerization of the silicate ions (e.g. quartz and feldspar) are more resistant to chemical weathering and form at later stages when magma crystallizes.

Each mineral consists of several major elements and is represented by a chemical formula. For example, quartz is SiO_2. Three types of feldspar, Na-feldspar ($NaAlSi_3O_8$), K-feldspar ($KAlSi_3O_8$) and Ca feldspar ($CaAl_2Si_2O_8$), are common. These feldspars form solid solutions. For example, the $NaAlSi_3O_8$–$KAlSi_3O_8$ solid solution is discontinuous, but the $CaAl_2Si_2O_8$–$NaAlSi_3O_8$ solid solution is continuous at high temperatures, and the Mg_2SiO_4–Fe_2SiO_4 solid solution (olivine) is continuous. The isomorphous substitution of ions in crystal structures depends on ionic radii, crystal structure, temperature, pressure, composition, and the thermochemical properties of the mother solution, e.g. magma or an aqueous solution. For example, the ionic radii of Fe^{2+} and Mg^{2+} are very similar, and so their substitution results in the formation of continuous solid solutions like Mg–Fe olivine.

The difference between the ionic radii of Na^+ and K^+ is large, so their substitution results in the formation of discontinuous solid solutions like Na–K feldspar. In crystals where bonding has a covalent character, isomorphous substitution is prevented.

The minerals other than silicate minerals mainly found near the surface environment include carbonate (e.g. $CaCO_3$, $CaMg(CO_3)_2$, $MgCO_3$, $FeCO_3$), sulfate (e.g. $CaSO_4$, $CaSO_4 \cdot 2H_2O$, $BaSO_4$, $SrSO_4$), sulfide (e.g. FeS_2, FeS, ZnS, PbS, $CuFeS_2$), oxide (e.g. Fe_2O_3, $Fe3O_4$), hydroxide (e.g. $Fe(OH)_3$, $Al(OH)_3$), and halide (e.g. $NaCl$, KCl). These minerals form from aqueous solution (e.g. seawater, hydrothermal solution) such as evaporate, limestone and hydrothermal ore deposits.

2.2.9.2 Rocks

Rock is classified according to its genesis and chemical and mineralogical compositions. From a genetical point of view, it is divided into igneous, metamorphic, and sedimentary rocks. Igneous rocks, which are volcanic, are classified into basalt, andesite, dacite, rhyolite, etc. based on chemical composition, mainly their SiO_2 contents by weight percent.

2.2.9.3 Igneous Rock and Igneous Activity

Igneous rock forms through the crystallization of molten magma, which is generally silicate melt. Minerals crystallize from the molten magma mainly due to changes in temperature and pressure. The chemical compositions of magma and minerals forming from it are different, with the crystallization process causing the changes. For example, the SiO_2 content of olivine, which forms early in magma's crystallization, is lower than that of the coexisting magma. Thus, the crystallization of olivine results in the enrichment of the SiO_2 content of magma compared with olivine. The SiO_2 content of magma increases as the crystallization process proceeds. After the crystallization of olivine, ferromagnesian minerals like pyroxene, biotite, and

Fig. 2.13 Phase diagram of binary (A, B) continuous solid solution system

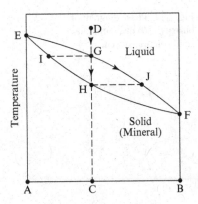

amphibole crystallize. In late stages of the crystallization process, feldspar (Na feldspar and K feldspar) and quartz crystallize.

The SiO_2 content of magma varies with the degree of crystallization. From SiO_2 content, igneous rock is classified into ultramafic rock (e.g. peridotite) that is about 40% by weight SiO_2, mafic rock (e.g. basalt) that is about 50% SiO_2, intermediate rock (e.g. andesite) that is about 60% SiO_2, and felsic rock (e.g. rhyolite) that is about 70% SiO_2. These rock types are subclassified depending on (1) the presence or absence of quartz, (2) the content ratio of feldspars, and (3) the kinds of mafic minerals present (Fe- and Mg-bearing minerals like olivine and pyroxene).

Next, the crystallization process during the cooling of magma is considered based on its phase diagram. The thermochemical stability fields of minerals are determined by temperature, pressure, and bulk chemical composition. A phase diagram of binary (A, B) solid solution systems is shown in Fig. 2.13. In this temperature-composition diagram, the solidus (line EIHF) and liquidus (line EGJF) show the chemical compositions of the solid and liquid phases, respectively, at equilibrium with liquid and solid forms at constant temperature, pressure and bulk composition. The assumed initial temperature and composition of magma are plotted as point D in Fig. 2.13. Decreasing temperature causes crystallization at G.

As the temperature decreases from G, the compositions of the solid and liquid phases change as $I \rightarrow H$, and $G \rightarrow J$, respectively. At H, the liquid phase disappears and only the solid phase exists. The composition of the solid phase (A/B ratio) at this point is AC:BC.

A solid solution whose composition changes continuously is termed continuous. An example such a solid solution is albite, or Na-feldspar ($NaAlSi_3O_8$) with anorthite, or Ca-feldspar ($CaAl_2Si_2O_8$) in solid solution. The anorthite component of this solid solution decreases with decreasing temperature.

A phase diagram showing the stability fields of minerals that do not form solid solutions is given in Fig. 2.14. At temperatures above line EIF, the liquid phase is stable. In the region encompassed by EIG and FIH, the liquid and solid phases are both stable. In the region GABH, solid phases A and B are stable. If the initial temperature and composition are plotted at D and the temperature decreases to J, the

Fig. 2.14 Phase diagram of
binary (A, B) discontinuous
solid solution system

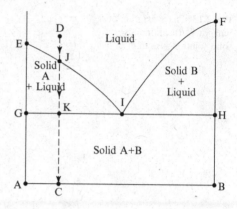

A phase crystallizes. Further decreasing the temperature results in the temperature-composition trend J → I. At I the liquid phase disappears and the solid phase B appears. As magma cools, there are discontinuous and continuous reaction sequences called Bowen's reaction series. The discontinuous reaction series proceeds from early to late stage crystallization from olivine to pyroxene to amphibole to mica to quartz and the continuous reaction series is Ca-rich feldspar (anorthite) to Na-rich feldspar (albite). Heavy minerals (e.g. magnetite, Fe_3O_4) crystallized from magma sink to the bottom of the magma chamber. In this case, crystals separate from magma in a process called fractionational crystallization. However, if a mineral crystallizes very slowly, it is called equilibrium crystallization. Strictly, crystallization does not occur in equilibrium conditions, but here equilibrium crystallization does not mean fractional crystallization. Fractional and equilibrium crystallizations are a source of the diversity of magmas and igneous rocks with different mineralogical and chemical compositions.

Magma composition changes as magmas are mixed and contaminated by host rocks. Primary basaltic magma is generated in the lower mantle. If mafic minerals such as olivine and pyroxene crystallize fractionally out from basaltic magma, the SiO_2 content of the basaltic magma increases. As the fractional crystallization proceeds, the magma becomes andesitic and finally dacite and rhyolite with high SiO_2 contents form. Although this process can produce felsic magma, most instances of felsic magma form by the melting of lower crust materials having high SiO_2 contents. Felsic magma formed in this way mixes with basaltic magma to form andesitic magma.

Different kinds of magma are formed by a variety of processes during the melting of source materials, the ascent of magma toward the surface and its solidification. The diverse compositions of magma are caused by the source rocks from which it is generated, the degree of partial melting of the source rocks, and the temperature, pressure and water content of magma.

For example, the water content of magma significantly influences its liquidus and solidus. Heat supplied to source rocks changes the degree of partial melting, resulting in different magma compositions. In the early stages of the study of magma genesis, fractional crystallization from primary magma (basaltic magma) was thought to be an important process causing the diversity of igneous rock. However, at present, other processes like magma mixing, contamination, the degree of partial

Fig. 2.15 Metamorphic reactions and stability field of some metamorphic minerals. Jd: jadeite, Qz: quartz, Na-feld: Na-feldspar, Ne: nephelline, Kya: kyanite, Sil: sillimanite, And: andalusite, Ara: aragonite, Cal: calcite

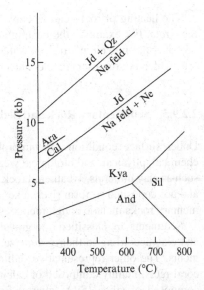

melting, and water content are thought to be the important processes and factors causing magma's variability.

2.2.9.4 Metamorphic Rock and Metamorphism

Minerals and rocks' textures change when their temperature and pressure conditions change. This process is called metamorphism. Rock that has suffered metamorphism is called metamorphic rock. For example, if the temperature and pressure of rock consisting of only Na feldspar stays at point B in Fig. 2.15 for a long time, the following reaction occurs.

$$NaAlSi_3O_8 \text{ (Na-feldspar)} \rightarrow NaAlSi_2O_6 \text{ (jadeite)} + SiO_2 \text{(quartz)}$$

Natural rocks, however, are more complicated and consist of multiple components and phases. Various metamorphic reactions occur simultaneously. Generally, chemical reactions proceed at higher rates as temperature and pressure increase. This kind of metamorphism is called progressive metamorphism. However, some metamorphism also occurs with decreasing temperature and pressure. This is called retrogressive metamorphism.

There are two types of metamorphism, the regional type and the contact type. Wide areas of rocks are metamorphosed by regional metamorphism. Regional metamorphism is now thought to be caused by the temperature and pressure changes that accompany plate subduction. These changes may depend on rate of plate subduction, the subduction slope, and the temperature and pressure distributions in the subduction zone etc. We can observe regional metamorphic rocks lying on the surface, implying that metamorphic rocks have been uplifted from deeper to shallower parts of the crust. The uplift mechanism is not well understood.

The heating of rocks causes contact metamorphism when magma intrudes into the strata. For example, the heating and recrystallization of limestone by magma, probably through circulating hot water, results in the formation of marble and horn-fels, which are representative contact metamorphic rocks.

2.2.9.5 Sedimentary Rock and Sedimentation

Under surface conditions, metamorphic, igneous, and sedimentary rocks suffer chemical, physical, and biological weathering, which produces secondary minerals such as clay minerals. Weathered rock fragments are transported by rivers and wind and are deposited to form river, lake, and seafloor sediments. Representative sedimentary rocks include conglomerate, sandstone, siltstone, and mudstone.

Sediments are classified into gravel, boulders, cobbles, pebbles, sand, silt, mud, and clay according to their grain size. The sedimentary rocks form from the sediments. The other representative sedimentary rocks are limestone from shells and coral reefs, mainly composed of calcium carbonate ($CaCO_3$, calcite), chert mainly composed of silica (SiO_2, quartz) from diatoms and radiolarian, and evaporite. Evaporite is formed by the evaporation of shallow seawater pools and ponds, resulting to the precipitation of salts (gypsum, $CaSO_4 \cdot 2H_2O$; anhydrite, $CaSO_4$; halite, NaCl; calcite, $CaCO_3$, etc.) Sediments being buried is associated with the dissolution and precipitation of minerals and compaction of the sediments, which reduces their porosity. This process is called diagenesis. Compact, solidified sediment is called sedimentary rock.

In earth's surface environment, mass transfer occurs continuously. This process changes the surface topography on a geologic time scale. The total amount of sedimentary rock is small compared to igneous rock, but it is very widely distributed.

Mass circulations associated with water and atmospheric circulations are driven by energy from the sun. The annual amount of energy in the earth's surface receives from the sun is 1.3×10^{24} calories excluding energy absorbed by the atmosphere. This is considerably larger than the energy from the earth's interior (Table 2.12).

Table 2.12 Various energy flow (Spilo et al. 1985)

Energy transfer process	Flux (cal/yr)
Total radiant energy from sun to universe	2.8×10^{33}
Solar energy input to earth	1.3×10^{21}
Solar energy influencing climate and biosphere	28.6×10^{23}
Energy for vaporization of water	2.2×10^{23}
Solar energy for photosynthesis	9.4×10^{20}
Energy for primary production	7.2×10^{20}
Energy from earth's interior to surface	3.1×10^{20}
Total fossil fuel energy used by humans (1975)	6.0×10^{19}
Total energy used for food by humans (1975)	3.2×10^{18}

Sedimentation and transportation processes are proceeding, caused by energy from the sun and from the earth's interior. Igneous activity and metamorphism occur, driven by energy derived only from earth's interior. Thus, these internal processes are distinctly different from external process such as weathering, erosion, transportation, and sedimentation, which are mainly caused by external energy from the sun. However, the rocks that are the source of material for igneous and metamorphic rocks are sometimes sedimentary rocks formed through the influence of external energy. Thus, the formation of igneous and metamorphic rocks is also influenced by external energy.

Weathering can be separated into two types, physical and chemical weathering. For example, temperature differences between day and night cause rocks to expand and shrink, which causes them to fracture and break down. This physical weathering does is not accompanied by chemical and mineralogical changes.

Chemical weathering is caused by chemical reactions between surface water like rainwater and rocks and involves chemical and mineralogical changes. Surface water contains CO_2 derived from atmospheric CO_2 and oxidation of organic matter in soil. Dissolution of CO_2 into water generates H^+ by the reaction

$$H_2O + CO_2 \rightarrow H_2CO_3$$

$$H_2CO_3 \rightarrow HCO_3^- + H^+$$

and

$$H_2CO_3 \rightarrow CO_3^{2-} + 2H^+$$

The pH of rainwater containing CO_2 in equilibrium with atmospheric CO_2 ($P_{CO2} = 10^{-3.5}$ atm) is calculated to be 5.65, and so rainwater is usually acidic. H^+ from the dissociation of H_2CO_3 and HCO_3^- reacts with minerals, dissolving cation from them. For example, silicate minerals react with H^+ according to the reaction:

silicate mineral $+ H^+ \rightarrow$ cation $+ SiO_2 +$ kaolinite (allophone (Al–Si–O–H amorphous phase), halloysite (metastable phase of kaolinite))

where the reaction coefficients are omitted for simplicity.

The order of solubility of silicate minerals in aqueous solution, calculated based on thermochemical data is consistent with Goldich's weathering series that gives the order of chemical weathering of natural rocks.

Other clay minerals such as smectite also form when silicate minerals dissolve. The formation of secondary minerals by chemical weathering depends on the degree of chemical reaction, the water/rock ratio, the type of aqueous solution, the rock and mineral species, etc.

When rocks weather, elements dissolve into aqueous solutions and are transported by rivers and ground water. Dissolved species such as ions and complexes, as well as colloids and fine particles are present in the aqueous solutions.

2.3   The Soils

Soils are distributed widely in the earth's surface environment. They are composed of various constituents including primary minerals in rocks, rock fragments, weathering products (secondary minerals like clays), and organic matter.

In addition to these solids, water, gaseous components, and organisms are found in soils. The weathering of rocks depends on rainfall, temperature, biological activity, the kinds and abundance of the rocks, and the amount and type of organic matter present. Climate and rock types are different in different regions, and so the properties of the world's soils vary a great deal, making classification of soils difficult. According to agricultural classification, there are the following types of soil: podsol, andosol, gley (gley lowland soil), dark red soil, red soil, and yellow soil. Forest soil includes podsol, brown forest soil, and black soil.

Soils are very important substances for agriculture, forestry, and mining, and thus we need to understand and improve the properties of soils and eliminate pollution of soils caused by these activities.

Soils used as mineral resources include bauxite for Al, and laterite for Fe and Ni. The bauxite is highly aluminous soil and is presently important Al ore. The laterite is red soil rich in Fe oxicles and/or Al.

2.4   The Biosphere

The biosphere is the sum total of all living and dead organisms on the earth. Components of the biosphere are found in every subsystem of the earth: surface water, both terrestrial water and seawater, the atmosphere, soils, and organisms. Recently, new species of microorganisms have been discovered deep underground, and on the seafloor and subseafloor. The total amount of organisms is called biomass. Organisms and other subsystems interact considerably. For example, photosynthetic reactions reduce atmospheric CO_2 and generate O_2. In the soil oxidation-reduction reactions (e.g. $Fe^{2+} + H^+ + 1/4O_2 = Fe^{3+} + 1/2H_2O$) occur through bacterial activity and significantly influence the geochemical behavior of elements including C, Mn, S, N, and V.

Photosynthetic reactions and the formation of carbonates by organic activity in aqueous solutions affect water quality.

Biological activity causes the formation of minerals. This process is called biomineralization. For example, coral and foraminifera make carbonates. Radiolaria and diatoms make shells of silica (SiO_2). Fish and mammals form bones and teeth composed of a phosphate mineral (apatite) inside their bodies. Acids produces by microorganisms dissolve silicate minerals. Life has evolved by interacting with the environment surrounding it, developing its remarkable diversity.

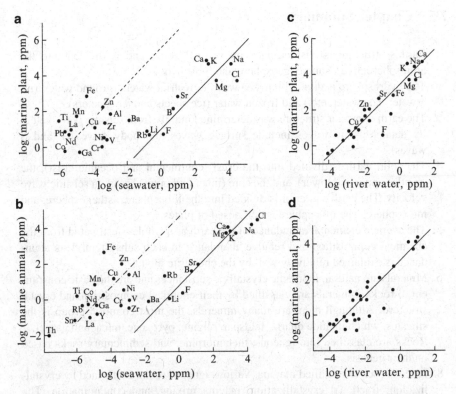

Fig. 2.16 The relationship between the chemical compositions of marine and terrestrial organisms and those of seawater and terrestrial waters (Nishimura 1991) (**a**) seawater vs marine plant, (**b**) seawater vs marine animal, (**c**) river water vs marine plant, (**d**) river water vs marine plant

The chemical composition of organisms is related to their environment. The relationship between the chemical compositions of marine and terrestrial organisms and seawater and terrestrial waters is shown in Fig. 2.16.

The properties and abundances of the constituents that make up the earth's surface environment, such as water, the atmosphere, soils, organisms, and humans, are very diverse. Ecological systems (ecosystems) in which each organism interacts with every other and exists in response to the flow of energy and mass can be classified according to the differences in their environments. The most basic subdivision is into terrestrial and aquatic systems.

The terrestrial ecosystem includes forests, meadows, deserts, and arable land. The aquatic system is subdivided into freshwater systems forests, rivers, lakes, and coasts and marine systems (open and coastal oceans). The characteristics of each ecological system, its diversity, the influence of human activity, and interactions with other subsystems are different.

2.5 Chapter Summary

1. Earth's structure is characterized by vertical zones and is divided into the atmosphere, hydrosphere, geosphere and biosphere.
2. The hydrosphere is divided into seawater, terrestrial waters (ground water, riverwater, lakewater, etc.) and frozen water (ice sheets and ice glaciers).
3. The earth's interior structure was determined mainly from seismic wave velocity data. Seismic waves include surface waves and body waves (P and S waves).
4. The solid earth is divided into the crust (continental and oceanic crust), the mantle (upper and lower), and the core (inner and outer) based on seismic wave velocity. The earth's interior is divided into the lithosphere, asthenosphere, and mesosphere. The lithosphere is composed of plates.
5. The average elemental abundance of the earth as a whole is estimated from the chemical composition and relative abundance in each subsystem. This abundance is explained relatively well by the chondrite model.
6. Minerals are natural, inorganic crystalline substances and are the basic constituent of rocks. Minerals are classified by their chemical composition and crystal structure. Although there are many minerals, the most common group is the silicates, which include quartz, feldspar, olivine, pyroxene, mica, and garnet.
7. Rocks are classified into igneous, metamorphic, and sedimentary rocks based on their genesis.
8. Igneous rock is solidified magma. Various igneous rocks are formed by crystallization, fractional crystallization, magma mixing, and contamination. The degree of partial melting of source rocks, the water contents of magma and the source rock, the temperature, and the pressure determine the compositions of magma and igneous rocks.
9. Metamorphic rocks form by metamorphism (changes in mineralogy, chemistry, and texture caused by changes in temperature and pressure) of igneous or sedimentary rocks. The important factors that produce different metamorphic rocks are temperature, pressure, the original rock, water, and CO_2.
10. Parental rocks are weathered, eroded, and transported by the action of surface water such as rain, ground water, or riverwater and the atmosphere. Materials transported mainly by rivers accumulate on the sea bottom as sediments (conglomerates, sand, silt, and mud). The sediments become buried, leading to diagenesis and solidification of the sediments. Solidified sediments are called sedimentary rock.
11. Soil occurs widely in earth's surface environment and is composed of primary minerals derived from original rock, secondary minerals such as clay minerals formed by weathering, and organic matter. It is a useful resource for obtaining metals such as Al, Fe, and Ni and for agriculture and forestry.
12. The biosphere exists in earth's surface environment, and it interacts with and influences other subsystems.

References

Anders E, Ebihara M (1982) Solar-system abundances of the elements. Geochim Cosmochim Acta 46:2363–2380

Ebihara M (2006) Chemistry of solar system, Shokobo (in Japanese)

Ernst WG (1969) Earth materials. Prentice-Hall, Englewood Cliffs

Ernst WG (ed) (2000) Earth system. Cambridge University Press, Cambridge

Goldich SS (1938) A study in rock weathering. J Geol 46:17–58

Holland HD (1978) Chemistry of the atmosphere and oceans. Wiley, New York

Holland HD, Petersen U (1995) Living dangerously. Princeton University Press, Princeton

Kawamura T, Iwaki H (eds) (1988) Environmental science I. Asakura Shoten, Tokyo (in Japanese)

Kobayashi J (1960) A chemical study of the average quality and characteristics of riverwaters of Japan. Ber. Ohara Inst-Landwirtschaft. Biol, vol 11. Okayama University, Okayama, p 313

Livingstone DA (1983) Data of geochemistry. In: Fleisher M, Fleisher M (eds) Data of geochemistry, 6th edn. USGS Prof. Pap, Washington, p 4400

Mason B (1958) Principles of geochemistry, 2nd edn. Wiley, New York

Matsuhisa Y, Akagi T (2005) General geochemistry. Baihukan, Tokyo (in Japanese)

McDonough WF (2004) Compositional model for the earth's core. In: Treatise on geochemistry, vol 2, The mantle and core. Elsevier, Amsterdam, pp 547–568

Palme H, O'Neil HSC (2004) Cosmochemical estimates of mantle composition. In: Treatise in geochemistry, vol 2, The mantle and core. Elsevier, Amsterdam, pp 1–38

Rudnick RL, Gao S (2004) Composition of the continental crust. In: Treatise on geochemistry, vol 3, The crust. Elsevier, Oxford, p 64

Spilo TG, Stigliani WM, Shoda M, Kobayashi T (translation) (1985) Science of Environment. Gakkai Syuppan Center, Tokyo (in Japanese)

Chapter 3
Material Circulation in the Earth

The earth system has fluid properties. Convection, circulation and fluid flow transport heat and materials continuously in the fluid and solid portions of the earth.

Materials are circulating in the earth's interior. It is thought that the mantle is convecting on a global scale. Hydrothermal solutions and seawater are circulating at midoceanic ridges and back arc basins at subduction zones. The terrestrial geothermal system circulates geothermal water that mainly originates from meteoric sources. Through convection and circulation, elements and materials such as rocks and minerals are transported locally and globally. The rates of movement of solid, liquid and gaseous materials differ considerably in different geologic processes. Earth's fluid character explains circulation near the earth's surface and in its interior via the hydrologic cycle, carbon cycle, solid circulation, plate tectonics, and plume tectonics.

Keywords Carbon cycle • Global geochemical cycle • Hydrologic cycle • Japanese island • Material circulation • Plume tectonics • Residence time • Subduction • Plate tectonics

3.1 Interactions in the Multispheres

Figure 3.1 shows the interactions between the multispheres in the earth system. Various materials and energies circulate within the system. The interactions occur not only between those subsystems that contact each other directly but also those that do so only indirectly. For example, plates generated at midoceanic ridges move laterally and subduct to the mantle. This is accompanied by magma generation deeper in the subduction zone. Magma then ascends to the upper part of the crust. The earth's surface environment is influenced considerably by the earth's interior activity as well as by external factors. For example, a meteorite impacts the earth's surface. If it is a

N. Shikazono, *Introduction to Earth and Planetary System Science*,
DOI 10.1007/978-4-431-54058-8_3, © Springer 2012

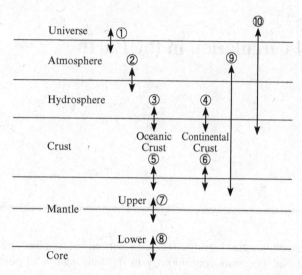

Fig. 3.1 The interactions in multispheres in the earth system. ① Formation of ozone layer. ② Dissolution of CO_2. ③ Weathering of basalt, circulation of hydrothermal solution, formation of hydrothermal ore deposits, sedimentation. ④ Ground water–soil–rock interaction, sedimentation, formation of weathering ore deposits. ⑤ Subduction of plate, metamorphism, igneous activity. ⑥ Igneous activity. ⑦ Mantle convection. ⑧ Core–mantle interaction. ⑨ Volcanic gas, igneous activity. ⑩ Meteorite impact

large body or bolide impact, mass extinction may result (Sect. 6.2.8). In the early stages of the earth's history, planetesimal and meteorite impacts on the earth's surface were very intense, and rapid degassing from the earth's interior occurred. This period is called the heavy bombardment period. Even at present, light gases like H_2 and He are being released from the earth's interior to the atmosphere.

The discussion above indicates that the earth is composed of several subsystems that interact, with inflows and outflows of materials and energy continuously and irreversibly occurring between subsystems. The various and complicated interactions occurring in earth's surface environment are considered in this book.

3.2   Hydrologic Cycle

Ocean water is heated by sun energy and evaporates, then falls as rain back into the ocean. The rates of annual water evaporation and rainfall are the nearly same in order of magnitude (Fig. 3.2). Inputs of water to the ocean from sources such as riverwater and rainfall and ocean output through evaporation are nearly balanced on a long-term scale. That means the amount of seawater is in a steady state. For the ocean as a whole, this cycle is nearly balanced. Locally, however, evaporation flux exceeds rainfall flux in arid regions. If evaporation proceeds considerably, salt

Fig. 3.2 The global hydrologic cycle, reservoirs of water (10^{15} kg and fluxes 10^{15} kg/year) (Natural Research Council 1993; Ernst 2000)

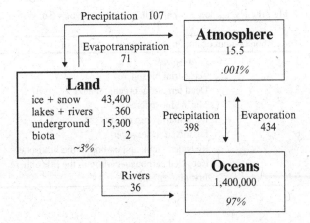

precipitates from evaporating seawater, resulting in the formation of evaporite (e.g. Dead Sea). Water moves not only between the atmosphere and the ocean, but also between the terrestrial environment, plants and the atmosphere. Figure 3.2 shows water circulation in the earth's surface environment.

Water circulates also globally in earth's interior system, including in the mantle. At midoceanic ridges, seawater circulates and reacts with oceanic crust, forming hydrous alteration minerals. The sediments above oceanic plates contain considerable amounts of seawater as interstitial water and hydrothermal solution originated from seawater. The hydrothermal solution reacts with oceanic crust, forming alteration minerals containing water. The oceanic crust, containing water, subducts with the plate. During the subduction, part of the water is released and moves upwards. Of the remaining water that subducts deeper, some is contained in magma and escapes as volcanic gas. Water generated by metamorphic dehydration reactions also moves upwards. It is known that most hydrous minerals decompose at depths shallower than 100 km. However, lawsonite ($CaAl_2Si_2O_7(OH)_2H_2O$) is stable below this depth. At high temperatures and pressures, lawsonite decomposes, releasing water. The water generated by the decomposition moves upwards and reacts with lehrzorite (a kind of ultramafic rocks) and brucite ($Mg(OH)_2$)-like rocks in the mantle wedge to form hydrous minerals such as serpentine. If serpentine moves to deeper parts, it may change to wadsleyite, a high pressure modification of olivine γ-Mg_2SiO_4. Thus, stable hydrous minerals and considerable amounts of water may be present in the mantle, and convection can carry some of the water upward upon its release from the decomposing hydrous minerals.

3.3   Carbon Cycle

It is important to consider the carbon cycle in the earth system for the following reasons: (1) increasing atmospheric CO_2 concentrations play an important role in global warming, (2) the carbon cycle has a significant affect on the energy cycle in

Table 3.1 The carbon content of important reservoirs (subsystems) (in units of 10^{15} g) (Holland 1978)

The carbon content of important reservoirs (in units of 10^{15} g)		
1	Atmosphere	690
2	Terrestrial biosphere	450
3	Dead terrestrial organic matter	700
4	Marine biosphere	7
5	Dead marine organic matter	3,000
6	Dissolved in seawater	40,000
7	Recycled elemental carbon in the lithosphere	20,000,000
8	Recycled carbonate carbon in the lithosphere	70,000,000
9	Juvenile carbon	90,000,000

the earth system, (3) carbon played important role in the origin of the atmosphere, seawater and life, and (4) carbon influences the cycles of other elements (S, O, etc.) in the earth system.

To consider the carbon cycle, we first need to know the amount of carbon stored in each reservoir or subsystem. Table 3.1 indicates that the amount of carbon in the geosphere is very large compared with other reservoirs like seawater, the atmosphere, and the biosphere. Fluxes between reservoirs are also summarized in Table 3.1 The dominant chemical state of carbon is different in different reservoirs. It is CO_2 in the atmosphere, organic carbon in the biosphere, HCO_3^- in seawater, and carbonates and carbon in the geosphere. Dominant reactions causing the transfer of carbon between reservoirs are (1) photosynthetic reactions, (2) carbon oxidation–reduction reactions, and (3) reactions between minerals and aqueous solutions. These reactions can be expressed as the following: photosynthesis, $CO_2 + H_2O \rightarrow CH_2O + O_2$; respiration decomposition of organic matter, $CH_2O + O_2 \rightarrow CO_2 + H_2O$; dissolution of calcite, $CaCO_3 + 2\,H^+ \rightarrow Ca^{2+} + CO_2 + H_2O$; dissolution of dolomite, $CaMg(CO_3)_2 + 2\,H^+ \rightarrow Ca^{2+} + Mg^{2+} + 2HCO_3^-$; dissolution of wollastonite, $CaSiO_3 + 2CO_2 + H_2O \rightarrow Ca^{2+} + 2HCO_3^- + SiO_2$; and dissolution of Mg-pyroxene (MgSiO), $MgSiO_3 + 2CO_2 + H_2O \rightarrow Mg^{2+} + 2HCO_3^- + SiO_2$.

Wollastonite is not common mineral. Ca-feldspar ($CaAl_2Si_2O_8$) is a more abundant Ca silicate mineral, but for simplicity of discussion, only the dissolution of wollastonite is listed above to represent Ca-silicate minerals. These dissolution reactions are caused by chemical weathering of silicate and carbonate rocks. HCO_3^- and cations, especially Ca^{2+}, Mg^{2+}, produced by these reactions are transported by ground water and rivers to the ocean and combine to form carbonates in reactions given by

$$Ca^{2+} + 2HCO_3^- \rightarrow CaCO_3\,(\text{calcite, aragonitee}) + CO_2 + H_2O$$

$$Mg^{2+} + 2HCO_3^- \rightarrow MgCO_3\,(\text{magnesite}) + CO_2 + H_2O$$

Combining these with the reactions accompanying the formation of carbonates and weathering reaction of silicates, we obtain

Table 3.2 The transfer rates of carbon between reservoirs (in units of 10^{15} g/year) (Holland 1978)

From	To	Process	(in 10^{15} g/year)
1	2	Net photosynthesis on land	48
2	1	Rapid decay of terrestrial organic matter	23
1	4	Net photosynthesis at sea	35
4	1	Rapid decay of marine organic matter	5
2	3	Accumulation of dead terrestrial organic matter	25
4	5	Accumulation of dead marine organic matter	30
3	1	Decay of dead terrestrial organic matter	25
5	1	Decay of dead marine organic matter	30

$$CaSiO_3 + CO_2 \rightarrow CaCO_3 + SiO_2$$

$$MgSiO_3 + CO_2 \rightarrow MgCO_3 + SiO_2$$

In these reactions, atmospheric CO_2 is transformed into carbonates. If carbonates settle onto the ocean floor, they subduct with the oceanic plate to deeper parts of the earth. CO_2 degasses by igneous and metamorphic reactions to the oceans and the atmosphere. This long-term carbon cycle is global and so is called the global carbon cycle.

As shown in Table 3.2, carbon fluxes from the geosphere to other reservoirs are small compared with other fluxes. Generally, the rate of inorganic reaction is smaller than that of organic and biogenic reactions. Most of the carbon in seawater and the atmosphere originated from the decomposition of carbon in dead organic matter. The carbon fluxes from the decomposition of organic matter and from the burning of fossil fuels including oil, coal, and natural gas are estimated to be 83×10^{15} and 4.2×10^{15} g/year, respectively. Therefore, the anthropogenic emission of carbon to the atmosphere is roughly 1/20 of the carbon flux due to the decomposition of organic matter. CO_2 in the atmosphere, transferred there by anthropogenic and natural processes is removed from the atmosphere and the hydrosphere to the biosphere by biological activity. Carbon is transferred to the atmosphere and the hydrosphere via weathering (Fig. 3.3) and also hydrothermal solutions, volcanic gasses, and gasses released by metamorphism. These fluxes are summarized in Fig. 3.3.

The carbon cycle relates to the oxygen cycle, which is obvious from the reaction

$$\text{``CH}_2\text{O''} + O_2 = CO_2 + H_2O$$

where "CH_2O" is used as shorthand notation to represent organic matter.

Therefore, it is obvious that the carbon cycle has a large influence on atmospheric O_2.

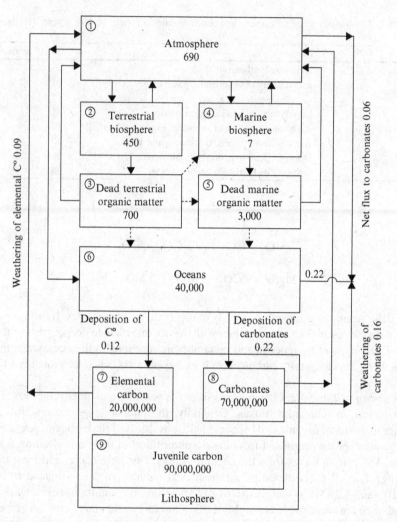

Fig. 3.3 A portion of the carbon cycle emphasizing C transfer during weathering and sedimentation (Holland 1978). Figure is carbon contents in units of 10^5 g of C and transfer rates in units of 10^{15} g of C/year

3.3.1 *Global Geochemical Cycle and Box Model*

As noted earlier, the earth system is composed of subsystems. These subsystems can be envisioned as boxes. Mass transfer occurring between the boxes is solved based on global material cycle models.

For simplicity, mass transfer between two boxes (Fig. 3.4) is considered below. Fluxes between two boxes (box 1 and box 2) are represented by

$$F_{12} = k_{12}M_1 \tag{3.1}$$

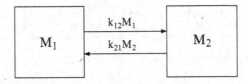

Fig. 3.4 A coupled two reservoir model system with fluxes proportional to the mass of the emitting reservoirs (Jacobsen et al. 2000)

$$F_{21} = k_{21}M_2 \tag{3.2}$$

where F_{12} and F_{21} are flux from box 1 to box 2 and from box 2 to box 1, respectively. M_1 and M_2 are the masses in box 1 and box 2, respectively, and k_{12} and k_{21} are rate constants for mass transfer from box 1 to box 2 and from box 2 to box 1, respectively.

Residence time (τ) is given by

$$\tau = M / F = 1 / k \tag{3.3}$$

The change of mass with time in boxes 1 and 2 is given by

$$dM_1 / dt = F_{21} - F_{12} = k_{21}M_2 - k_{12}M_1 \tag{3.4}$$

$$dM_1 / dt = F_{12} - F_{21} = k_{21}M_2 - k_{12}M_1 \tag{3.5}$$

If mass does not change with time, the steady state condition is given by

$$dM_1 / dt = dM_2 / dt = F_{21} - F_{12} = k_{21}M_2 - k_{12}M_1 = 0 \tag{3.6}$$

In this case, $F_{21} = F_{12}$ and $k_{12}M_1 = k_{21}M_2$.
Next, we consider the nonsteady state where

$$M_1 + M_2 = M = constant \tag{3.7}$$

This condition means no input from other boxes to box 1 or box 2 and no output from box 1 or box 2 to other boxes. From (3.4), (3.5), and (3.7), we obtain

$$M_1 = M_{10} + k_{21}(M_{12} + M_{21}) / (k_{12} + k_{21}) - \{(k_{12}M_{12} - k_{21}M_{20})$$
$$/(k_{12} + k_{21})\} \exp\{-(k_{12} + k_{21})t\} \tag{3.8}$$

$$M_2 = M_{20} + k_{12}(M_{12} + M_{21}) / (k_{12} + k_{21}) - \{(k_{21}M_{20} - k_{12}M_{10})$$
$$/(k_{12} + k_{21})\} \exp\{-(k_{21} + k_{12})t\} \tag{3.9}$$

where M_{10} and M_{20} are equal to M_1 and M_2 (t=0), respectively.

Equations 3.8 and 3.9 indicate that the change of mass with time beginning at t=0 depends on the response time, $1/(k_{12} + k_{21})$. The responses of reservoirs (boxes) that undergo perturbations (rapid flux increases or decreases) can be determined

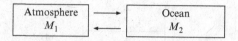

Fig. 3.5 A coupled two reservoir (seawater, atmosphere) system

based on (3.8) and (3.9). In a two-box model, the differential equations can be solved easily if the values of the parameters k and M_0 are known. However, a multibox model is difficult to solve based on analytical methods. Numerical and matrix solution methods are useful for determining changes of mass over time. Examples of applications of these methods to the global geochemical cycle can be found in Berner et al. (1983), a BLAG model; and Berner (1994), which are GEOCARB models, Chameides and Perdue (1997) and Kashiwagi and Shikazono (2003).

Rapid increases in human activity have caused a perturbation in the CO_2 concentration of the atmosphere. We consider the response of the atmosphere to this perturbation and the change of mass of carbon in the biosphere.

The changes in M_1, the amount of carbon in the atmosphere (box 1), and M_2, the amount of carbon in the biosphere (box 2) with time are given by (Lasaga 1981) as

$$M_1 = 690 + 50 \exp(-0.11015t) \tag{3.10}$$

$$M_2 = 450 - 50 \exp(-0.11015t) \tag{3.11}$$

where t is time in years and the values of initial mass are $M_{10} = 740$ and $M_{20} = 400$.

Ten years later, the amounts of carbon in the atmosphere and the biosphere are calculated to be 707×10^5 and 433×10^5 g, respectively.

Carbon in the atmosphere moves into seawater, as well as into the biosphere. The interaction between the atmosphere and seawater will be considered below.

Here, the atmosphere and seawater are regarded as box 1 and box 2, respectively (Fig. 3.5).

Using the amount of carbon in box 1 and box 2 and fluxes from box 1 to box 2 and from box 2 to box 1 (80 Gt/year), the rate constant, k, can be estimated. Using estimated k values, we can calculate changes in the amount of carbon in the atmosphere and seawater over time (Fig. 3.6). In this case, the amount of carbon in the atmosphere and seawater settles at a constant value (a steady state condition) ten thousand years later.

The exchange reaction for carbon between seawater and the atmosphere considered above is $CO_2 + H_2O = HCO_3^- + H^+$. However, in addition to this reaction, there are reactions in seawater as follows.

$$Ca^{2+} + 2HCO_3^- = CaCO_3 + CO_2 + H_2O \tag{3.12}$$

$$HCO_3^- + H^+ = CH_2O + O_2 \tag{3.13}$$

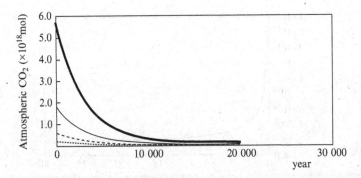

Fig. 3.6 Atmospheric CO_2 variation with time for atmosphere–seawater system (Shikazono 2007). *Thick curve* is for the amount of present-day atmospheric $CO_2 = 5.5 \times 10^{16}$ mol and oceanic $CO_2 = 3.3 \times 10^{20}$ mol. The other curves are for the different initial amounts of atmospheric and oceanic CO_2

These reactions are caused by biological activity. If we take these reactions into account, a three-box model has to be used to represent the atmosphere, seawater, and the biosphere.

Next, let us consider the carbon cycle in the seawater–atmosphere (fluid earth)–crust (solid earth) system in which the response of each subsystem, particularly the solid earth, is very slow. Here, the atmosphere and seawater can be modeled as one box by assuming that equilibrium between the atmosphere and seawater has been attained. This assumption is reasonable because the reaction rate between the atmosphere and seawater is very rapid. Although the solid earth consists of the crust, mantle and core, the mantle and core are not considered in this model.

Changes in the amount of carbon in the two boxes (the crust and seawater + atmosphere) over time were calculated using values of rate constant (k) estimated from the amounts of carbon in the boxes and the fluxes today (Fig. 3.7). We assumed that k does not change with time and the atmosphere and seawater are in equilibrium with respect to CO_2.

Figure 3.7 show the changes of the amounts of carbon in the crust over time for the last 30 Ma. This shows it takes 10^5 years to attainment a steady state condition. This suggests that the geosphere (crust) does not play an important role in the efficient removal of atmospheric CO_2 produced by human activity.

The calculations shown above seem to be useful in making a first approximation of the response of reservoirs to an external perturbation. However, the following problems remain.

1. The real natural system is more complicated than a two-box system, and consists of multiple boxes.
2. The results of calculations based on the two-box model cannot be evaluated by comparing with the changes in the real system.
3. k is not constant but rather variable with time.

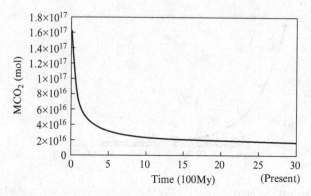

Fig. 3.7 Carbon cycle between atmosphere. ocean (fluid earth) and crust (solid earth) (Shikazono 2007). M_1, M_2 and M_{Ca} (Ca mol in ocean) is assumed to be $1,000 \times 10^{22}$ mol, $33,875 \times 10^{18}$ mol and V_s(volume of ocean $= 1.32 \times 10^{21}$ l) $\times 0.01$ mol, respectively

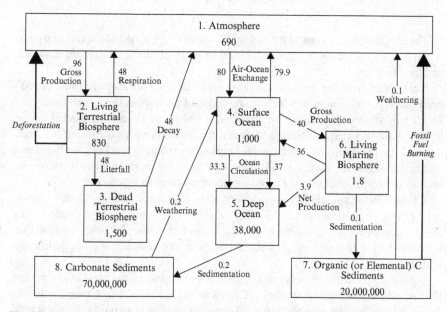

Fig. 3.8 The preindustrial, steady-state global cycle for C using eight-box model (Chameides and Perdue 1997). The *boxes* represent specific reservoirs, with the *numbers inside the boxes* being reservoir amounts in units of Gtons C (1 Gton C $= 1 \times 10^9$tons $= 1 \times 10^{15}$ g) and the *arrows* fluxes between reservoirs with the numbers having units of Gtons C/year. The *heavy arrows* denote the pathways by which anthropogenic activities perturb the global cycle

4. Flux from human activity into the atmosphere is assumed to be simultaneous, but it should be expressed as a function of time.

5. We assumed that F is linearly correlated with M, but this is uncertain.

Chameides and Perdue (1997) performed model calculations taking into account problems 1–4. They considered seven boxes: the atmosphere, terrestrial living organisms, the surface seawater, deep seawater, living marine organisms, organic matter in sediments, and carbonate sediments.

According to the multibox model, the change of mass with time is expressed as

$$dM_i(t)/dt = \sum_{j=1}^{N} K_{ij}M_j(t) \quad (i = 1,2,\ldots,N)$$ (3.14)

$$K_{ij} = k_{j \to i(i \neq j)}$$

$$K_{ij} = -\left(\sum_{j \neq i}^{N} k_{ij}\right)$$

Assuming eight boxes in Fig. 3.8 and steady state conditions, we obtain

$$dM_1/dt = F_{21} + F_{31} + F_{41} + F_{71} - F_{12} - F_{14} = 0$$
$$dM_2/dt = F_{12} - F_{21} - F_{23} = 0$$
$$\vdots$$ (3.15)

We can obtain k values from (3.15). Then, using initial values of mass, assuming constant values of k and using the differential matrix equation, we can calculate how M changes with time (Chameides and Perdue 1997).

Figure 3.9 shows the results for the case where CO_2 input from humans to the atmosphere changes rapidly in 1920. This calculation does not agree with records of the change of atmospheric CO_2 over time. The most serious problem with this calculation method is that this does not take into account time-dependent k. In contrast, using a pseudo-nonlinear model that assumes that k is variable with time produces results that agree fairly well with the historical data. The pseudo-nonlinear model, then, is more useful for predicting atmospheric CO_2 concentration in the future. It has its own problems, however, including the following:

1. Predicting future atmospheric CO_2 concentration is difficult because we are not very certain of parameter values and
2. The assumption of $F = kM$ is difficult to apply to a real system.

Changes in atmospheric CO_2 concentration depend on many anthropogenic factors like economics, ethics, policy, science, and technology. We do not discuss these issues here, but limit our consideration to the models and their applicability. The pseudo-nonlinear box model is based on the equation $F = kM$ where k is a function of time. However, in real system, the nonlinear equation $F = kM^n$ seems more appropriate. We need to know reaction coefficient (n) value and the rate-limiting mechanism to use this model.

For example, the flux of carbon from the left to right side of the equation for photosynthetic reactions ($CO_2 + H_2O = CH_2O + O_2$) is given by Holland (1978) as

$$F = a\left\{1 - \exp(-bP_{CO2})\right\} - c$$ (3.16)

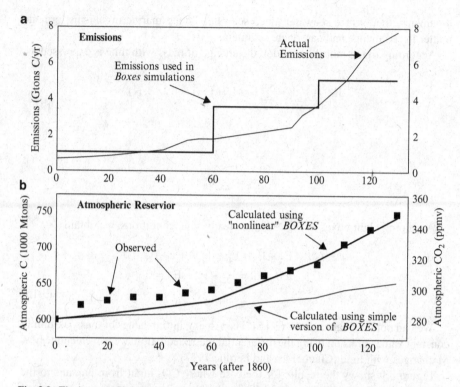

Fig. 3.9 The impact of human activities in the global C cycle since 1860. (**a**) Anthropogenic CO_2 emissions since 1860. (**b**) The atmospheric C reservoir (left-hand axis) and CO_2 mixing ratio (right hand axis) as a function of year since 1860 (Chameides and Perdue 1997)

F does not correlate linearly with P_{CO2} (a,b: constant, c: respiration rate in the absence of CO_2).

Next, we consider the mass transfer between the geosphere and the hydrosphere. The transfer of carbon between the crust and seawater is given by the reaction

$$CaSiO_3 + CO_2 = CaCO_3 + SiO_2 \tag{3.17}$$

Equation 3.17 is a simplified reaction and can be derived from the following two reactions.

$$CaSiO_3 + 2CO_2 + 2H_2O = 2HCO_3^- + Ca^{2+} + SiO_2 \tag{3.18}$$

$$Ca^{2+} + 2HCO_3^- = CaCO_3 + CO_2 + H_2O \tag{3.19}$$

Atmospheric CO_2 is fixed as $CaCO_3$ through reactions (3.18) and (3.19) for weathering, precipitation of $CaCO_3$ in the ocean and sedimentation on the seafloor.

In order to estimate the carbon flux using these reactions, the reaction rates in (3.18) and (3.19) must be estimated. These reaction rates do not have a simple linear relation with P_{CO_2}. The general equation expressing dissolution and precipitation reactions of solid phases is given by

$$dC/dt = k(\sigma^m - 1)^n \qquad (3.20)$$

where C is concentration, σ is the saturation index, equal to I.A.P./K_{eq} (I.A.P is the ion activity product, and K_{eq} is the equilibrium constant), k is apparent reaction rate constant, and m and n are constant coefficients.

This equation means that the reaction rate (dC/dt) does not depend on concentration in a linear relationship. In one component system, the precipitation rate is represented by

$$dC/dt = -k\{(C/C_{eq})^m - 1\}^n \qquad (3.21)$$

where C_{eq} is the equilibrium concentration.

Here, we assume that the reaction occurs irreversibly.

In the case of m = 1 and n = 1, (3.21) is converted into

$$dC/dt = -k(C - C_{eq}) \qquad (3.22)$$

If $C \gg C_{eq}$ (far from equilibrium conditions), we obtain

$$dC/dt = -kC \qquad (3.23)$$

This equation indicates that the flux decreases linearly with time.

The relationship between V, C and M is given by

$$VC = M \qquad (3.24)$$

where V is volume of the system and M is the mass of the water.

If V is constant, $d(M/V)/dt = -k(M/V)$ and $dM/dt = -kM$.

This indicates M decreases linearly with time. Therefore, a linear relation, F = kM, is established if n = 1, m = 1 and the system is far from equilibrium. In this case, a pseudo-nonlinear multibox model is useful for deriving the relation between mass and time. The pseudo-nonlinear model has been successfully used to simulate the biogeochemical cycles of C, S, P, N, and O (Chameides and Perdue 1997).

A global sulfur cycle model and flux is shown in Fig. 3.10 and Table 3.3, respectively. The change in the amounts of S in the soils and the atmosphere with time after rapid addition of anthropogenic S by influences the preindustrial global S cycle, which had been in steady state.

It is noteworthy that the time needed to attain steady state is very short, about 30 years, about the same as for the N cycle, but shorter than the carbon cycle, which takes more than 100 years.

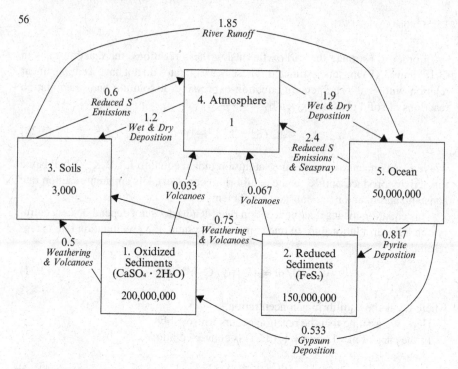

Fig. 3.10 Global sulfur cycle (Chameides and Perdue 1997)

Table 3.3 Preindustrial fluxes of sulfur for the five-steady reservoir steady state model (Chameides and Perdue 1997)

Fluxes	Flux (Tmoles year^{-1})
From oxidized S sediments to soils ($F_{1\rightarrow3}$) to atmosphere ($F_{1\rightarrow4}$)	Weathering = 0.4
	Volcanoes = 0.1
	Total = 0.5
	Volcanoes = 0.033
From reduced S sediments to soils ($F_{2\rightarrow3}$) to atmosphere ($F_{2\rightarrow4}$)	Weathering = 0.6
	Volcanoes = 0.15
	Total = 7.5
	Volcanoes = 0.067
From soils to atmosphere ($F_{3\rightarrow4}$) to ocean ($F_{3\rightarrow5}$)	Reduced S emissions = 0.6
	River runoff = 0.85
From atmosphere to soils ($F_{4\rightarrow3}$) to ocean ($F_{4\rightarrow5}$)	Wet and dry deposition = 1.1
	Seasalt deposition = 0.1
	Total = 1.2
	Wet and dry deposition = 0.6
	Seasalt deposition = 1.3
	Total = 1.9
From ocean to atmosphere ($F_{5\rightarrow4}$) to oxidized sediments ($F_{5\rightarrow1}$) to reduced sediments ($F_{5\rightarrow2}$)	Reduced S emissions = 1.0
	Seaspray = 1.4
	Total = 2.4
	Gypsum deposition = 0.533
	Pyrite deposition = 0.817

Fig. 3.11 The cycles of Hg
at preindustrial (**a**) and
postindustrial (**b**) stages
(Garrels et al. 1975)

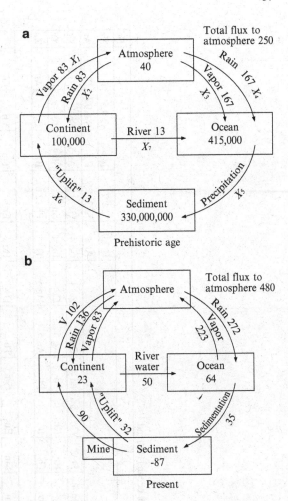

The anthropogenic influences on other elements' global cycles could be studied,
but few such works have been carried out. The cycles of toxic base metals (Hg, Mn,
Pb, etc.) in the pre- and postindustrial stages have been studied (Fig. 3.11) (Garrels
et al. 1975), but no computations on the cycles are available.

3.3.1.1 Factors Controlling the Chemical Composition of Seawater

Inputs from riverwater to the ocean and outputs from the ocean by sedimentation
are important factors controlling the chemical composition of seawater. Residence
time (τ) of an element in seawater is defined as

$$\tau = A_0 / (dA_{R,O} / dt) \tag{3.25}$$

Table 3.4 Residence time (log τ, in year) of elements in ocean (Holland 1978)

H 4.5																	He
Li 6.3	Be (2)											B 7.0	C 4.9	N 6.3	O 4.5	F 5.7	Ne
Na 7.7	Mg 7.0											Al 2	Si 3.8	P 4	S 6.9	Cl 7.9	Ar
K 6.7	Ca 5.9	Sc 4.6	Ti 4	V 5	Cr 3	Mn 4	Fe 2	Co 4.5	Ni 4	Cu 4	Zn 4	Ga 4	Ge	As 5	Se 4	Br 8	Kr
Rb 6.4	Sr 6.6	Y	Zr 5	Nb	Mo 5	Tc	Ru	Rh	Pd	Ag 5	Cd 4.7	In	Sn	Sb 4	Te	I 6	Xe
Cs 5.8	Ba 4.5	La 6.3	Hf	Ta	W	Re	Os	Ir	Pt	Au 5	Hg 5	Tl	Pb (2.6)	Bi	Po	At	Rn
Fr	Ra 6.6	Ac															

Ce	Pr	Nd	Pm	Sm	Eu	Gd	Tb	Dy	Ho	Er	Tm	Yb	Lu
Th (2)	Pa	U 6.4											

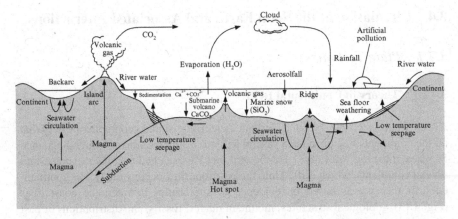

Fig. 3.12 Various processes controlling chemical composition of seawater (Shikazono 1992)

where A_0 is the total amount of a given element in seawater and $dA_{R,O}/dt$ is the flux of the element from riverwater to the ocean. The τ of various elements can be calculated because the values for $dA_{R,O}/dt$ and A_0 are known (Table 3.4). Residence times vary widely in a range of $10^2–10^8$ years. One important factor controlling τ is the chemical state of the element in seawater. Elements with large τ values such as alkali earth, alkaline elements, and halogens are stable as ions and complexes in seawater. Base metal elements with small τ values tend to be present as fine particles and colloids. These particles and colloids settle onto the seafloor in a short time.

Seawater's chemical composition is determined not only by riverine input and sedimentation, but also by evaporation, weathering of seafloor basalt, formation of evaporites, aerosol fall, biological activity, circulation of seawater and hydrothermal activity at midoceanic ridges, island arcs and back arcs, volcanic gas inputs, seepage from seafloor sediments near subduction zones and continental coasts, and mass transfer between bottom seawater and interstitial water in seafloor sediments and anthropogenic pollution (Fig. 3.12). The most important factor controlling the concentration is different for different elements. For example, the removal of Mg from seawater through the circulation of seawater and hydrothermal solutions below the seafloor, such as happens at midoceanic ridges, has a considerable effect on seawater's Mg concentration. Alkali elements such as Li and K and base metal elements such as Cu and Zn are released from the oceanic crust by hydrothermal solutions into seawater. Potassium is removed from seawater and precipitated in clay minerals in weathered basalt. Ca transported by hydrothermal solutions to seawater precipitates as carbonate through biological activity. Carbonate dissolves in deep seawater. The seawater depth below where carbonate dissolves is called the CCD (carbon compensation depth). Silica in seawater is enriched into the shells of diatoms and radiolaria. Dead diatoms settle onto the seafloor, resulting in their being the main component of biogenic SiO_2 in sediments. The marine snow settling onto the sea bottom adsorbs dissolved species and colloids and adsorbed matter can also desorb during settling. Thus, we are certain that biological activity is very important for controlling seawater chemistry.

3.4   Circulation of the Solid Earth and Associated Interactions

3.4.1   Plate Tectonics

3.4.1.1   Theory of Continental Drift

Alfred Wegener (1880–1930), a German meteorologist and geographer posited that all the continents had once been assembled into a supercontinent, which he named Pangaea, meaning "all Earth". He thought that Pangaea had become separated into several continents, which drifted into the positions they occupy today. His continental drift theory was based on evidence of similarities between the coast lines (Fig. 3.13), geological structures, mountain ranges, fossils, and distributions of past climate zones, as witnessed by the presence of glaciers coal, or rock salt, of two continents that are very distant from each other today. However, many scientists, particularly geophysicists, rejected this idea because it did not include any physical mechanism that could drive the motion of the continents. But in the 1950s, paleomagnetic data supporting this view became available. Paleomagnetic characteristics of many rock samples from many locations in various continents clearly indicated that if the magnetic poles have been fixed, the continents have been drifting.

3.4.1.2   Ocean Floor Spreading

In 1950s, mountain ranges called ridges, which run long distances (totaling about 60,000 km) under the center of many oceans, were discovered on the seafloor. Detailed investigation revealed ridge valleys, or rifts, in the middle of these ridges. Rifts are observed on the continents. Figure 3.14 shows a cross section of a rift valley in east Africa. The topography of this rift valley is notably similar to that of the midatlantic rift. Deep lakes like Lake Tanganyika are distributed along the rift valley.

Many measurements revealed a strong correlation between seafloor topography and continental heat flow. Heat flow at ridges is high compared with other areas. This suggests that heat convection may be occurring in the mantle, causing mantle convection, as proposed by Arthur Holmes (1890–1965) (Fig. 3.15) and continental drift. Peleomagnetic evidence strongly supports mantle convection theory. As shown in Fig. 3.16, magnetic anomalies are distributed as nearly straight lines parallel to the ridge axis. When magma intrudes at ridges, it cools and solidifies. Beyond the Curie point of about 570°C, the rocks are magnetized in the direction of earth's magnetic field at the time of cooling. The recorder model, called the Vine–Mathews hypothesis can reasonably explain measured magnetic anomaly data (Vine and Mathews 1963). Paleomagnetic study has clearly indicated a continuing history of ocean floor spreading.

Ages of ocean floor sediments were obtained from radiometric age and microfossil data. The ages obtained by these methods and magnetic anomaly data showed that the ocean floor gets older as its distance from the ridge increases (Fig. 3.17); the

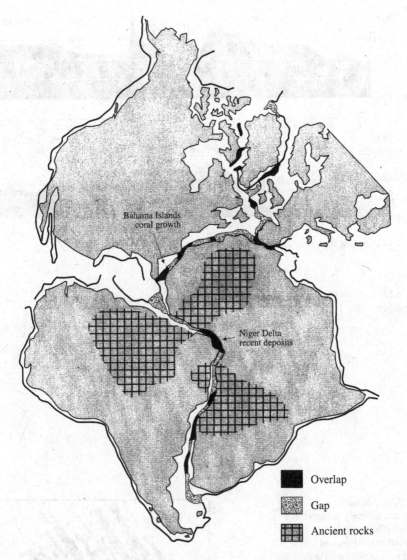

Fig. 3.13 The jigsaw-puzzle fit of continents bordering the Atlantic ocean formed on the basis of Alfred Wegener's theory of continental drift (Bullard et al. 1965; Holland and Petersen 1995)

age of ocean floor is generally younger than the continents; and the oldest ocean floor is from the Jurassic period. These pieces of evidence lead to the idea of ocean floor spreading and renewal.

3.4.1.3 Plate Tectonics

According to plate tectonics the earth's solid surface is divided into more than ten separate rigid blocks or plates (Fig. 3.18). They are moving laterally and interact

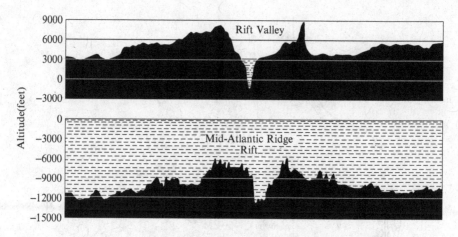

Fig. 3.14 Cross section of rift valley, Africa and Mid-Atlantic ridge (Holmes 1978)

Fig. 3.15 Cartoon of mantle convection

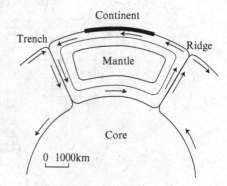

Fig. 3.16 Measured magnetic anomaly pattern for a small portion of the Mid-Atlantic Ridge just southwest of Iceland. Positive anomalies are shown in *black*; negative anomalies in *white* (Heirtzler et al. 1966; Ernst 2000)

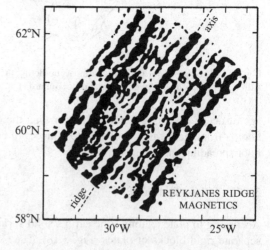

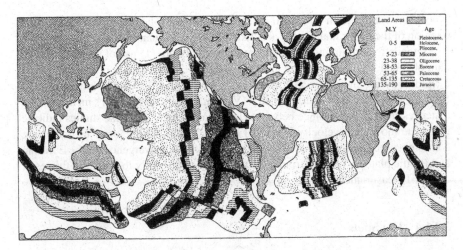

Fig. 3.17 Age of the ocean floor based on magnetic anomaly patterns (Pitman et al. 1974; Holland and Petersen 1995)

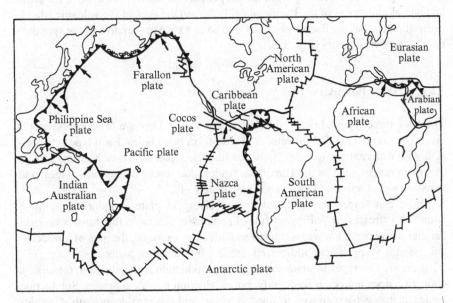

Fig. 3.18 Major lithospheric plates and their boundaries as deduced from worldwide seismicity tectonics (Dewey 1980; Ernst 2000)

with each other, causing the characteristic features of geological structures observed in the crust like mountain building, folding, faulting, etc.

The plates include both the crust and upper mantle. A low-velocity zone was found in the upper mantle by seismological investigation. The plates are considered to be the section above this zone, which is called the lithosphere. The thickness of the lithosphere ranges from 60 to 200 km. Below the lithosphere is the asthenosphere, a region of the upper mantle that acts like a fluid rather than a solid.

Methods for estimating plate movement velocity are (1) use of ocean floor age data obtained by paleomagnetic and other means, (2) calculation of the rotation of a rigid body around an axis of rotation (Euler pole), (3) study of hot spots, and (4) very long base line interferometry (VLBI), which can precisely measure long distances such as the distance between Hawaii and Japan. All these studies indicate that plate movement velocity is several cm year^{-1} and their velocity estimates are consistent with each other. For example, the velocity of convergence for the Eurasian and Pacific plates estimated from rotating velocities is nearly the same as the result of VLBI measurement.

3.4.1.4   Hot Spots

As shown in Fig. 3.19, the age of the islands of the Emperor Sea mount and Hawaiian Island Chain is older toward the northwest. This trend can be explained by the combination of magma ascending from the deeper part of mantle over the last 10^8 years and lateral movement of the Pacific plate. Based on the ages of the volcanic islands it is estimated that the Pacific plate has moved at 8 cm/year relative to hot spot during the last 6×10^5 years (Fig. 3.20).

3.4.1.5   Plate Boundary

There are three types of plate boundaries (Fig. 3.21). They are (1) convergent, (2) divergent, and (3) parallel movement or transform boundaries. Each type is associated with different topography. Trenches like the Japan trench or Mariana trench, mountain ranges such as the Himalaya, Andes, and Rocky Mountains, ridges, and transforms characterize (1), (2), and (3), respectively.

Mid-ocean ridges are representative of divergent plate boundaries. At these boundaries stresses are pulling apart the lithosphere. The faults that form boundary-parallel margins are known as transform faults. For example, the axis of a midoceanic ridge is vertically cut and moved laterally by transform faults.

There are two types of plate convergence: subduction and obduction. Generally, obduction does not occur frequently, but subduction is very common. Subduction means that one plate moves to another plate, sinking into deep earth's interior (mantle) at a convergent plate boundary, while obduction is one model of the plate convergence, that oceanic crust overrides or overthrusts onto the leading edge of a continental lithospheric plate. In obduction, a part of the oceanic crust known as an ophiolite is thrust onto the continental crust. Studies on ophiolites can give us information on the structure and composition of the ancient oceanic crust. An ophiolite is a section of the earth's oceanic crust and consists of sediments, basaltic pillow lava, sheeted dikes, gabbro, cumulate peridotite, and tectonized peridotite, in descending order.

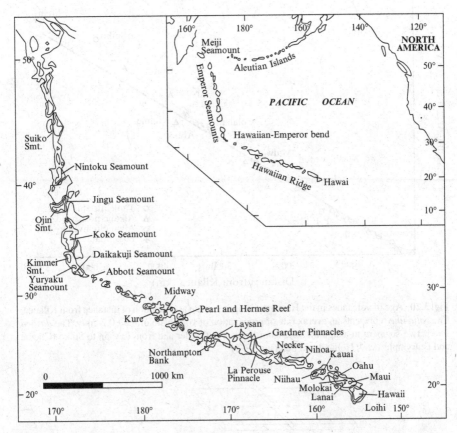

Fig. 3.19 Bathymetry of the Hawaiian-Emperor volcanic chain (Holland and Petersen 1995). Contours at 1 and 2 km depths are shown in the area of the chain only. The *inset* shows the location of the chain (outlined by 2 km depth contour) in the central North Pacific (Clague and Dalrymple 1989)

In deeper parts of subduction zones such as in the area of the Japanese archipelago or the Andes Mountains, magma is generated and ascends to form intrusive bodies of igneous rocks like granite, and regional metamorphism occurs. In subduction zones high mountains are built. For example, the Indian continent, on the Indian and Australian plates, collided with the Eurasian plate and subducted underneath it, resulting in the building of the Himalayan Mountains.

3.4.1.6 Plate Convergent Boundary

Figure 3.22 is an outline index map of the Japanese subduction zone. In the figure, thick lines with trenches are converging plate boundaries and arrows indicate the relative plate motions.

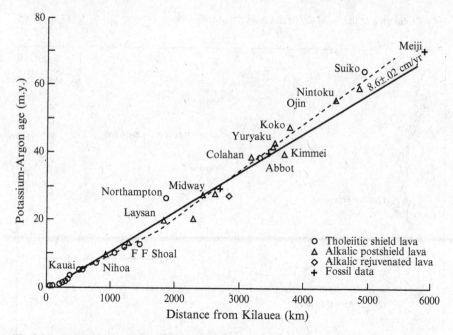

Fig. 3.20 Age of volcanoes in the Hawaiian-Emperor chain as a function of distance from Kilauea. The *solid line* represents an averax rate of propagation of volcanism of 8.6±0.2 cm/y. The *dashed line* is two-segment fit a sign the date from Kilavea to Gaeder and from Laysan to Suiko (Clague and Dalrymple 1989; Holland and Petersen 1995)

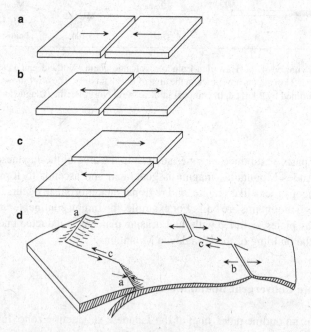

Fig. 3.21 Three types of plate boundary (Sugimura et al. 1988). (**a**) convergence, (**b**) divergence, (**c**) parallel movement (transform fault), (**d**) transform fault links

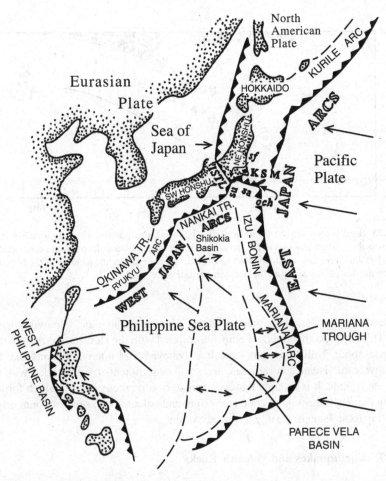

Fig. 3.22 Outline index map of the Japanese subduction zones (Uyeda, 1991). *Thick lines with teeth* are converging plate boundaries. *Arrows* indicate relative plates motions. Abbreviations: su, Suruge trough; su, sagami trough; sf, south Fossa Magna triple junctions; Och, Off Central Honshu triple junctions; ISTL, Itoigawa-Shizuoka Tectonic Line; KSML, Kashima VLBI station (Uyeda, 1991)

A deep trench, known as the Japan trench, runs parallel to northeastern Japanese Island on the Pacific side (Fig. 3.23). The Japan trench is more than 10^4 m deep in some places. It was formed by the subduction of the Pacific plate. In the Japanese Islands, there are high mountains like the Japanese Alps and rapid uplift is observed in mountainous areas. This uplift and mountain building were caused by the compression of the Eurasian plate and North American plate by the Pacific plate and Philippine Sea plate. Numerous dykes run in the strata with many trending nearly E–W. This indicates that Pacific plate is moving nearly E–W.

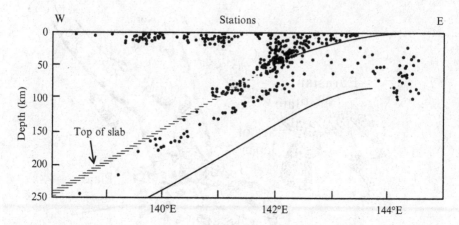

Fig. 3.23 Section showing the focal depth distribution of small earthquakes along an east–west line across the northeastern part of the Japan arc and the double-planed deep seismic zone detected by the Tohoku University seismic network. The hatched zone shores the position of the boundary between the descending slab and the overlying mantle (Bott 1982)

The Philippine Sea plate, which collides with the Asian plate, is moving nearly N–S. This suggests that the Izu Peninsula collided with the Honshu, the main island of Japan about 3 million years ago. It is believed that microcontinents like Izu Island, volcanic islands, coral reefs, and seafloor sediments have collided with the Japanese Islands. It is thought that widespread occurrences of chert, which formed deep in the sea, limestone of coral reef origin and volcanic rocks of seamount origin in the Japanese Islands are due to this accretion.

3.4.1.7   Earthquakes and Volcanic Rocks

Earthquakes in subduction zones occur deep along a plane, called a Wadachi–Benioff plane. The cross section of northeast Honshu, Japan clearly indicates how the plate subducts and earthquakes and volcanisms occur along this plane (Figs. 3.23 and 3.24). Volcanic zones are distributed parallel to Japan trench and Nankai trough where Pacific plate and Philippine Sea plate subduct, respectively.

3.4.1.8   Subduction Mode

Uyeda and Kanamori (1979) summarized the slopes of Wadati–Benioff zones (zones of active seismicity) in subduction zones around the world (Fig. 3.25).

The slope of the subduction at Mariana is the steepest, while the slab subducting with the shallowest slope is under Chile. Uyeda and Kanamori (1979) classified the mode of subduction into Mariana- and Chilean-types (Fig. 3.26). According to their study, the features of geologic events associated with these two modes of subduction are distinct. For example, a back arc basin exists

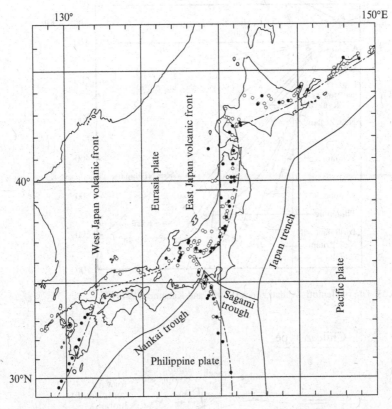

Fig. 3.24 Distribution of volcano, volcanic front and plate boundaries in Japanese Islands and surrounding area (modified after Sugimura et al. 1988). *Solid circle*: active volcano, *Open circle*: the other Quaternary volcano

between the continent and the island arc in Mariana-type subductions, but not in Chilean-type subduction zones. Big earthquakes like those in Chile in 1960, Alaska in 1964 tend be associated with Chilean-type but not Mariana-type subduction. One possible explanation for the two distinct modes of subduction is the different ages of plates. Young plates have relatively high temperatures and low densities. Thus, young plates like the Chilean plate subduct at a gentle slope, causing a compressional stress regime, generation of big earthquakes, intermediate-type (andesite) volcanism and associated ore deposits like porphyry copper-type deposits. In contrast, older plates subduct with steep slopes, causing extensional stress regimes and forming backarc basins, bimodal mafic (basaltic)–felsic (dacitic) volcanism and associated ores known as Kuroko deposits which are massive stratabound polymetallic deposits formed on the seafloor in sub-seafloor environments (Sect. 4.3.4).

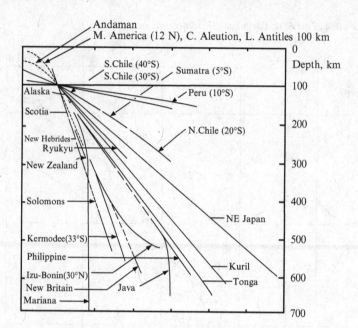

Fig. 3.25 Dip of Benioff–Wadati zones (Uyeda and Kanamori 1979)

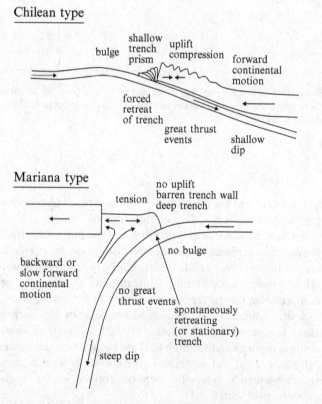

Fig. 3.26 Two modes of subduction boundaries (Uyeda and Kanamori 1979)

3.4.1.9   Geologic History of the Japanese Islands

The geologic history of the Japanese Islands can be explained in terms of plate tectonics.

The Japanese Islands are thought to have been formed mainly by accretion. They are composed mostly of accretional prisms from the Permian, Jurassic and Cretaceous ages including chert, limestone, sandstone, basalt, etc. The Japanese Islands were originally located at the eastern margin of the Asian continent. Paleomagnetic study has revealed from 20 to 15 million years ago, the Sea of Japan opened rapidly and the Japanese Islands separated from the Asian continent and moved to their present-day position.

In the Tertiary (Miocene) age, intense submarine volcanism occurred on the seafloor of the Sea of Japan accompanied by the deposition of Green tuff, a thick, altered pile of volcanic materials associated with mudstone and the formation of Kuroko deposits. Northeast Japan (Honshu) uplifted starting in the late Miocene and its terrestrial area expanded. In the Quaternary age volcanism occurred related to subduction of the Pacific plate and the Philippine Sea plate.

In the Cretaceous age intense igneous activity occurred. Granitic rocks formed at that time. Granitic rocks are widely distributed in the Japanese Islands. This is due to uplift and erosion after the Cretaceous age.

Regional metamorphic rocks like those found in the Sanbagawa metamorphic belt, which is composed of high pressure type metamorphic rock, and Hida metamorphic complex, which is low pressure type are distributed as zones in the Japanese Islands. The Hida metamorphic rocks are old, up to 2 Ga, indicating that they have been a part of the Asian continent and then separated from it.

3.4.2   Plume Tectonics

The use of earthquake tomography revealed the three-dimensional structure of the earth's interior. P wave velocity increases with increasing density, and density depends on temperature. Thus, using earthquake tomography, we can estimate the distribution of temperature in the earth's interior in three dimensions. This method revealed that high- and low-temperature zones exist in the interior (Figs. 3.27 and 3.28). This heterogeneous distribution of temperature is explained by the presence of hot plumes and cold plumes. It has been posited that hot plume activity causes hot spot volcanism and volcanism at midoceanic ridges. If hot plume activity intensified, volcanic activity at midoceanic ridges, as well as oceanic crust production rates and thus subduction rates, would increase, resulting in magma generation, degassing from magma, an increase in the atmospheric CO_2 concentration, and global warming.

Plume tectonics is the theory that plumes govern mantle dynamics. Plume tectonics may control plate tectonics, meaning that plume tectonics has considerable effects on both earth's surface and interior systems.

Fig. 3.27 Tomographic image of the earth's mantle beneath the Japanese arc, down to the core/mantle boundary showing the distribution of slow and fast seismic waves. The wave velocity distribution also reflects temperature distribution and shows the penetration of a cold subducting slab through the transition zone into the lower mantle (Rollinson 2007)

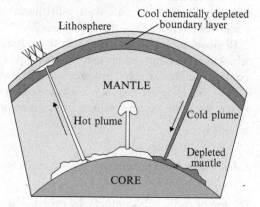

Fig. 3.28 The Cambell and Griffiths (1993) plume-driven mantle convection model. (**a**) The pre-4.0 Ga earth with descending, depleted cold plumes, and (**b**) the 2.0–0 Ga earth with ascending enriched plumes (Rollison 2007)

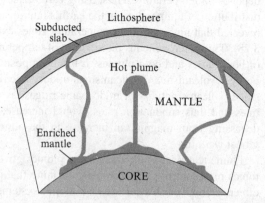

One possible explanation for the generation of hot and cold plumes is that subducting slabs stay at the boundary between the upper and lower mantles and these "stagnant slabs" fall as cold plumes deeper into the mantle, resulting in the formation of hot plumes as hotter mantle material ascends.

Scientists think that igneous activity in earth-type planets (Mercury, Mars, and Venus in our solar system) is caused by plume tectonics. However, no evidence of plate tectonics is available for these planets. Thus, it is likely that subduction of plates and stagnant slabs is not what generates hot and cold plumes there.

3.5 Chapter Summary

1. In the earth system, circulation of fluids such as the atmosphere and water and solids including plate movement, plume movement, and mantle convection are occurring.
2. Geochemical cycles of various elements and climate changes can be elucidated using computer simulations based on multibox models.
3. Geological processes in the earth's surface environment such as mountain building, earthquakes, faulting, folding, volcanism, and metamorphism can be explained in terms of plate tectonics.
4. The dynamics and physical properties of the earth's interior can be explained in terms of plume tectonics. Plume tectonics may control plate tectonics and the earth's surface environment in terms of long-term climate change and cause igneous activity in the earth-type planets Mercury, Mars, and Venus.

References

Berner RA, Lasaga AC, Garrels RM (1983) The carbonate-silicate geochemical cycle and its effects on atmospheric carbon dioxide over the past 100 million years. Am J Sci 283:641–683

Berner RA (1994) GEOCARB II—A revised model for atmospheric CO_2 over Phanerozoic time. Am J Sci 294:56–91

Bott MHP (1982) The interior of the Earth: its structure, constitution and evolution. Elsevier, New York

Bullard EC, Everett JE, Smith AG (1965) The Fit of the continents around the Atlantic. Phil Tras Roy Soc Lond 1088:41–45

Cambell IH, Griffiths RW (1993) The evolution of the mantle's chemical structure. Lithos 30:389–399

Chameides P (1997) Biogeochemical cycles. Oxford University Press, New York

Clague DA, Dalrymple GB (1989) Tectonics, geochronology and origin of the Hawaiian-Emperor volcanic chain. In: Winterer EL, Huss DM, Decker RW (eds) The Eastern Pacific Ocean and Hawaii, vol N, The geology of North America. Geological Society of America, Boulder, pp 188–217

Dewey JL (1980) Episodicity, sequence and style at convergent plate boundaries. In: RW Strangway (ed) The continental crust and its mineral resources. Geological Association of Canada, Special Paper 20:553–574

Ernst WG (2000) Earth systems. Cambridge University Press, Cambridge

Garrels RM et al (1975) Chemical cycles and the global environment—assessing human influences. William Kaufmann, Los Altos

Heirtzler JR, LePichon X, Baron JG (1966) Magnetic anomalies on the Reykjanes Range. Deep Sea Res 13:427 (Abstract)

Holland HD (1978) The chemistry of the atmosphere and oceans. Wiley, New York

Holland HD, Petersen U (1995) Living dangerously. Princeton University Press, Princeton

Holmes A (1978) Principles of physical geology, 3rd edn. Nelson, London

Jacobsen MC, Charlson RI, Rogue H, Orians GH (eds.) (2000) Earth system science. International Geophysics Series, vol 72. Elsevier Academic Press, Amsterdam

Lasaga AC (1981) Dynamic treatment of geochemical cycles. In: Lasaga AC, Kirkpatrick RJ (eds) Kinetics of geochemical processes: reviews in mineralogy, vol 8. Mineralogical Society of America, Washington, pp 69–110

Natural Research Council (1993) Solid-Earth sciences and society. Natural Academy of Sciences, Washington, DC

Rollison H (2007) Early Earth system. Blackwell, Oxford

Sugimura A, Nakamura Y, Ida Y (eds) (1988) Illustrated Earth science. Iwanamishoten, Tokyo (in Japanese)

Uyeda S (1991) The new view of the Earth. Freeman, San Francisco

Uyeda S, Kanamori H (1979) Subduction of oceanic plate and formation of marginal sea. Kagaku 48:91–102 (in Japanese)

Vine FJ, Mathews DH (1963) Magnetic anomalies over oceanic ridges. Nature 199:947–949

Chapter 4
Nature–Human Interaction

In previous chapters, we described non-human subsystems and the interactions between them. In this chapter, we undertake the interactions between humans and natural non-human subsystems including (1) natural disasters, (2) earth's environmental problems and (3) natural resources.

Keywords Acid rain • Destruction of ozone layer • Earthquake • Energy resources • Global warming • Mineral resources • Natural disasters • Natural resources • Nature–humans interaction • Water resources

4.1 Interactions Between Nature and Humans

Figure 4.1 shows the different features of interactions between the natural and human systems in which materials and heat flow. Humans take natural resources from earth's surface environment. These natural resources are consumed by human society, which also generates wastes. Most of the wastes emitted by humans are stored in the natural system. The stored waste changes the earth's surface environment, causing environmental problems. There are both direct and indirect environmental problems. In direct problems, wastes directly influence humans. For example, drinking water that has been contaminated by human activity is a direct environmental problem. An example of an indirect environmental problem is global warming due to an increase in atmospheric CO_2 emitted by human activity. In this case, CO_2 emitted by human activity does not directly influence humans, but the temperature increase caused by higher atmospheric CO_2 causes changes in earth's environment, thus influencing humans. If an environmental problem is global, we call it one of earth's environmental problems, while if it is local, it is a waste problem and/or environmental pollution. Waste emitted from humans to natural systems pollutes the environment, then returns back to humans and influences them directly and indirectly. This sort of feedback system characterizes earth's environmental problems.

N. Shikazono, *Introduction to Earth and Planetary System Science*,
DOI 10.1007/978-4-431-54058-8_4, © Springer 2012

Fig. 4.1 Earth environmental problem associated with mass and energy flow between humans and nature

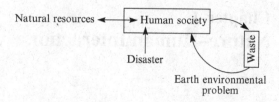

In contrast, natural disasters are characterized by a one-way flow of energy and matter from the natural system to humans. This flow is not continuous but intermittent with respect to time and is generally not global but local.

Humans cannot control the generation or scope of natural disasters. It is expected that with the development of scientific research and technology, we can predict when and where natural disasters occur and prevent them in the future.

4.2   Natural Disasters

Natural disasters are natural phenomenon causing damage to humans. The characteristic features of natural disasters are as follows:

1. There are various kinds of natural phenomena, and so there are also various kinds of natural disasters.
2. Secondary disasters are often precipitated by primary disasters. Examples of secondary disasters are fires, tsunamis, and mudflows associated with earthquakes and volcanic eruptions.
3. The features and types of natural disasters change over time. For example, natural disasters like earthquakes have come to cause more intense damage to human society with the development of civilization.

4.2.1   Classification of Natural Disasters

Natural disasters are classified into external and internal forms based on the driving force and natural phenomenon involved. Exogenic energy comes from outside of the earth, mainly being energy from the sun, while endogenic energy comes from the earth's interior.

4.2.1.1   External Force Natural Disasters

Climate disasters such as those caused by wind, snow, rain, drought, crop damage from cold weather, mist, etc; seawater disasters involving waves and tidal waves and riverwater disasters (floods) are caused by external forces.

4.2.1.2   Internal Force Natural Disasters

Volcanic disasters like lava or pyroclastic flows and ground disasters like landslides, mud floods, liquefaction of sandy ground, and earthquakes are caused mainly by internal forces. The main driving force for landslides, mud floods and floods is gravity energy. Water plays an important role in causing these phenomena. The water cycle is mainly driven by external energy from the sun. Thus, it could be said that external energy also causes these phenomena indirectly.

Climate disasters are also caused by atmospheric circulation driven mainly by external energy from the sun. Topography also has a significant influence on atmospheric circulation patterns. The balance of erosion, uplift, sedimentation, and subsidence combine to form topography. Those processes are in turn caused by both internal and external forces. Thus, internal force is also important as a cause of quick-onset climate disasters. In addition to this sort of natural disasters, human society suffers long-term disasters. Representative long-term disasters are sea level change, desertification, and El Nino. El Nino is a phenomenon where seawater along the coast of Ecuador and northern Peru becomes warmer by several degrees in winter.

In some cases, natural disasters are caused by both anthropogenic and natural effects and it is not easy to decide which effect is dominant. Landslides are sometimes considered to be natural disasters caused by anthropogenic activity such as development of building estates and deforesting.

4.2.2   Earthquakes

Earthquake disasters are characterized by widespread damage and frequent large-scale generation of secondary disasters. For example, tidal waves and landslides often happen associated with earthquakes. Subduction of oceanic plates causes accumulation of regional stress energy in continental plates, resulting in faulting if the stress exceeds the maximum strength of rock with respect to fracturing. If the seafloor is uplifted by faulting due to an earthquake, tidal wave can be generated that can travel a very long distance.

For example, the tidal waves generated by the huge magnitude 9.5 Chilean earthquake (23 May 1960) and magnitude 9.0 Japanese earthquake (11 March 2011) caused extensive damage to the south coast of Hokkaido and the east coast of Iwate, Miyagi, Fukushima, and Ibaraki prefectures in Japan.

There is a tendency for huge earthquakes to occur on the east coast of the Pacific region but not on the west coast (Fig. 4.2). This is thought to be because the regions have different modes of subduction. Generally, huge earthquake generation is associated with Chilean-type subduction but not with Mariana-type (Fig. 3.27).

The borders of the Pacific plate, Philippine Sea plate, Eurasian plate, and North American plate all lie in the vicinity of the Japanese Islands. The distribution of earthquakes that occur and mechanisms that generate them can be explained by

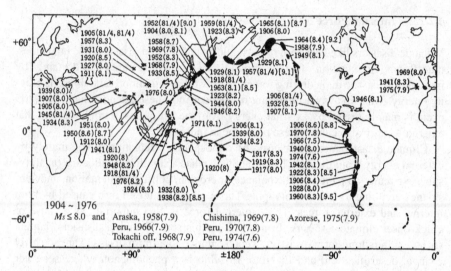

Fig. 4.2 Huge earthquake during 1904–1976 (Kanamori 1978). 1904–1976 Ms > 8.0 and Araska 1958(7.9), Chishima 1969(7.8), Azores 1975(7.9), Peru 1966(7.9), Peru 1970(7.8), Tokachi off 1968(7.9)

considering their interactions. Earthquakes occurring near the Japanese Islands are classified into the following types: plate boundary, inland shallow, and subducting oceanic plate interior. For example, the Hanshin earthquake in 1995 in Japan was an inland-type related to the movement of an active fault. In addition to subduction-related earthquakes, magma-related earthquakes can happen. Earthquakes associated with magma intrusion at midoceanic ridges and island arcs are this type. Generally, shallow-type earthquake and earthquakes that happen in a series are not huge. Examples of series earthquakes in Japan are the Matsushiro earthquake and the Izu earthquake.

4.2.3 Volcanic Disasters

Volcanic disasters are caused by volcanic sedimentation and crustal movement associated with volcanic eruption. Examples are disasters caused by volcanic eruptions, pyroclastic flows, lava flows, mud flows, landslides, and volcanic gas emissions.

Pyroclastic flow means the high-speed flow of a very hot mixture of pyroclastics and volcanic gases on the surface. This can cause serious disasters. This phenomenon is associated with felsic-andesitic volcanism. The basaltic magma flows with relatively high speed due to its low viscosity compared with felsic and andesitic magma. Due to this property, basaltic lava flows cause disasters such as the collapse of houses as in Maunaloa, Hawaii. The pyroclastic flow associated with felsic and intermediate (dacite and andesite) volcanism causes intense disasters such as those that occurred at the Unzen-fugen and Asama volcanoes in Japan.

If volcanic ash and volcanic gas reach the stratosphere, climate changes like a decrease in temperature can result. For example, the eruption of Pinatubo volcano in the Philippines resulted in global cooling. More than 1,700 people died due to the large amounts of volcanic gas containing CO_2 that erupted from Lake Nyos, Cameroon. SO_2 and HCl gases from volcanic gas that enters riverwater enhances its acidity, lowering its pH. Changes in pH and toxic base metal concentrations can influence riverwater ecosystems. The characteristics of volcanic disasters mentioned above can be summarized as follows:

1. Volcanic eruptions are not continuous but intermittent. There are many volcanoes in the world, but few are active.
2. Volcanoes are distributed in restricted areas. More than 60% of the volcanoes in the world are concentrated in the Circum-Pacific belt. Active volcanoes occur in subduction zones. In the Japanese Islands, located in a subduction zone, there are many volcanoes distributed in volcanic zones, but the area where volcanism occurs and the area suffering volcanic disasters are very limited, lying only near the volcanic eruption center.
3. Generally, volcanic disasters are not as destructive as earthquake disasters. However, exceptionally huge volcanic eruptions and pyroclastic flows associated with them can cause disasters over a wide area. The biggest historic volcanic eruption damaged an area several hundreds of kilometers in radius. In another example, the volcanic eruption of Sakurajima in southern Kyushu, Japan more than 12,000 years ago was its largest, and its volcanic ash has been found as far as Aomori, northern Japan, 1,100 km away.

4.3 Natural Resources

4.3.1 Renewable and Nonrenewable Resources

When metallic ores are exhausted from a mine, we can no longer obtain metals from that site. These types of resource is called nonrenewable. Minerals and fossil fuels like oil, coal, and natural gas are nonrenewable resources, while biomass and water are renewable. In general, however, it is not easy to distinguish between nonrenewable and renewable resources. Recently, hydrothermal ventings and formations of ore deposits on the deep seafloor were discovered. These submarine hydrothermal ore deposits develop very rapidly. Thus, it seems likely that substantial amounts of new hydrothermal ore could form after mining of existing submarine hydrothermal ore deposits begins. In this case, we could not classify these hydrothermal ore deposits as nonrenewable resources. In a long-term geological process, new mineral resources and fossil fuels may form if humans mined out currently existing resources. However, the short-term recovery of these resources by humans is impossible. We can control the character, quantity, and distribution of biomass, but two of the

Fig. 4.3 Consumption rate of total metal with time (Meadows et al. 1992)

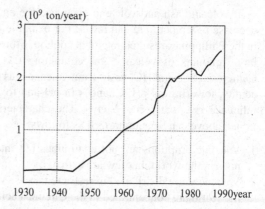

characteristics of mineral resources and fossil fuel energy resources are that (1) we cannot control these features, and (2) we cannot obtain the same resources from the same place after they are mined out.

4.3.2   Resources Problems

Recently, we have faced resources problems such as (1) the extraction, refinement, and consumption of resources have increased rapidly, and (2) humans have used up much of the known energy and mineral resources. The rate at which humans use energy and mineral resources has been growing over time. A rapid increase in the rate of consumption of total metals over time is shown in Fig. 4.3. The causes of this rapid increase are considered to be an increase in the worldwide population, the use of metals in various ways, and the development of technology. These increases in the extraction and consumption of resources may result in their serious depletion. The other resources problem is the uneven distribution of resources across the world. For example, most of the chromium (Cr) ore in the world is produced in the Republic of South Africa. Many metal mines have been operated in Japan; however, only a few metal mines are operating today and Japan imports most of the metallic ores it uses. Figure 4.4 shows amount of aluminum (Al) resources consumed on each continent. It clearly shows that Al consumption per capita in Asia is small, irrespective of Asia's large population.

It is generally said that our main resources problem is energy. However, there are various natural resources besides energy, among these minerals. The reasons mineral resources problems are not seen as a major part of the general resources problem are that we have generally been able to find alternatives to use in place of particular minerals and recycling of mineral resources is sometimes possible. In contrast, the recycling of fossil fuels or the energy derived from them is not possible, and alternatives to fossil fuels are not widely used yet. Large amounts of gas,

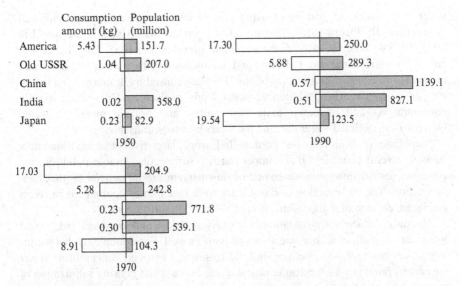

Fig. 4.4 Resources consumption amount on each continent (Nishiyama 1993)

especially CO_2, are produced in the use of fossil fuels. These gasses are difficult to fix, although some disposal technologies such as underground CO_2 sequestration are being developed (Sect. 4.4.6).

4.3.3 Exploitation of Natural Resources and Associated Environmental Problems

As mentioned above, humans use a variety of resources. In exploiting natural resources we damage the environment. For example, we have caused a global decrease in biomass due to deforestation. The annual mass decrease of tropical forests in Asia, Africa, and South Africa is about 18×10^5, 13×10^5, and 41.2×10^5 ha, respectively. The main causes of this decrease are slash-and-burn agriculture and over-pasturage. For example, 70% of the loss of tropical forests in Africa is due to slash-and-burn agriculture. The forest mass reduction caused by slash-and-burn agriculture is related to decreased productivity of the land over the last 40 years, corresponding to rapid population growth.

Currently, the amount of water people use is increasing. The rate of increase in various fields like agriculture, mining, and industry is higher than the population growth rate. Generally, water resources are thought to be renewable, but recently we have been using so much that it is becoming nonrenewable. For example, we have used large amounts of ground water, resulting in a lowering of ground water levels. If the amount of ground water in reservoirs decreases, we turn to using more surface water from rivers etc. instead of using ground water, resulting in environmental destruction. To illustrate this change, runoff from the main rivers in the world in 1971–1975 was only 17–40% of the amount of runoff before 1955. This change was caused by agricultural use. Pollution and acidification of riverwater and lakewater are also serious environmental problems. They are caused by drainage from industry and acid rain. The pollution of water leads to decreased dissolved oxygen concentrations and increased toxic base metals and organic matter in water. Organisms in polluted water die, and the water becomes unusable.

Sometimes, pollution crosses borders. In Europe, large rivers like the Rhine flow through several countries. If the upper reaches suffer considerable pollution, the countries further down the watershed of downstream are influenced by that contamination. The construction of dams leads to destruction of ecosystems in rivers and lakes, erosion of soils, salinization of soils and desertification.

Not only surface water contamination of rivers, lakes, and rainwater, but ground water contamination has become a serious issue as well. For example, global warming causes sea levels to rise, pushing the freshwater (ground water)/(brine water (seawater) boundary underground near the coast inland and causing salinization of the ground water there.

4.3.4 Mineral Resources

Mineral resources include all the minerals that are useful to humans. They are divided into metallic and nonmetallic types.

4.3.4.1 Metallic Mineral Resources

Metallic mineral resources are taken from metallic ore deposits in which useful metallic elements are enriched. The degree of enrichment of metal elements in ore deposits is represented by their economical concentration factor, defined as the ratio of the concentration of metals in the leanest minable ore to their concentration in average crustal rocks (Skinner 1976). This ratio differs for different elements (Fig. 4.5, Table 4.1). Generally, the concentration factor for elements that are common in the crust ("common elements") such as Al and Fe is low, but for elements that are rare in the crust ("rare elements") such as Hg, W, or Au, the ratio is high (Fig. 4.4).

4.3.4.2 Classification and Genesis of Metallic Ore Deposits

Metallic ore deposits are classified according to their genesis: (1) magmatic ore deposits, (2) hydrothermal ore deposits, (3) sedimentary ore deposits, and (4) weathering ore deposits.

Fig. 4.5 Economical
concentration factors of some
commercially important
elements (Skinner and Porter
1987). Economical
concentration
factor = abundance in ore
deposit divided by crustal
abundance

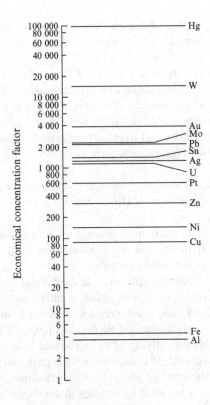

Table 4.1 Economical concentration factors of some commercially important elements (Siever et al. 2003)

Element	Crustal abundance (wt%)	Economical concentration factor[a]
Aluminum	8.00	3–4
Iron	5.8	5–10
Copper	0.0058	80–100
Nickel	0.0072	150
Zinc	0.0082	300
Uranium	0.00016	1,200
Lead	0.00010	2,000
Gold	0.0000002	4,000
Mercury	0.000002	100,000

[a]Economical concentration factor = abundance in ore deposit divided by crustal abundance

Magmatic Ore Deposits

Magmatic ore deposits are formed by the enrichment of minerals containing useful metallic elements that become separated out from magma. The separation of sulfide melt from silicate melt and the sinking of heavy minerals (e.g. magnetite (Fe_3O_4)) in magma chambers due to gravity cause the enrichment of certain elements. Representative magmatic deposits are Ni, Cu, Fe, Cr, and Pt deposits associated with ultramafic rocks and Ti deposits associated with anorthosite. Ni tends to occur as sulfides (e.g. NiS) and Fe and Cr as oxides (Fe_3O_4 and Cr_2O_3).

In late stages of crystallization of felsic magma, silicate melt forms that contains a large amount of water and CO_2. Uncommon elements like rare earth elements are enriched in late-stage felsic magma, from which pegmatite forms. Carbonatite is rock formed from carbonate melt. Thus, the rare earth elements Ti, Nb and P are enriched in carbonatite ore.

Hydrothermal Ore Deposits

A hydrothermal solution is a high temperature (100–600°C) aqueous solution. Hydrothermal ore deposits form from hydrothermal solutions via precipitation of useful metal elements.

An example of a hydrothermal ore deposit formed at high temperatures is porphyry copper, which is characterized by dispersion of Cu sulfides in granitic rocks and is associated with Mo and Au.

Hydrothermal ore deposits form at 200–350°C at midoceanic ridges and backarc basins. Mixing of hydrothermal solutions and cold seawater near the seafloor is considered to be a main formation mechanism. Seawater penetrates deeper below the seafloor and is heated by magma. Heated seawater interacts with host rocks, accompanied by changes in the chemical composition and hydrothermal alteration of the host rocks. Heated modified seawater, as a hydrothermal solution, contains considerable amounts of base metals released from host rocks, and ascends rapidly to the seafloor through fractures and fissures. Base metals (Cu, Pb, Zn, Fe, etc.), precious metals (Au, Ag) and rare metals (Mo, As, Sb, etc.) in the hydrothermal solution precipitate near the seafloor due to mixing with cold seawater. Figure 4.6 shows a hydrothermal solution circulation system associated with midoceanic ridge hydrothermal ore deposits. The hydrothermal system consists of a heat source (magma), rocks, and the hydrothermal solution. Heat and mass are transported by the hydrothermal solution. Precipitation of minerals onto the seafloor forms chimneys (Fig. 4.7). Black smokers containing fine particles of sulfides (chalcopyrite ($CuFeS_2$), pyrrhotite ($Fe_{1-x}S$), pyrite (FeS_2), sphalerite (ZnS), galena (PbS), etc.) and white smokers containing white minerals such as barite ($BaSO_4$), anhydrite ($CaSO_4$), and silica (SiO_2) issue from chimneys (Fig. 4.6). Since the first discovery of black smokers and chimneys 2,600 m deep at 21°N East Pacific Rise in 1979, many hydrothermal ore deposits have been found at midoceanic ridges and back arc basins (Fig. 4.8).

Metals enriched into back arc deposits such as in Okinawa, Mariana, Fiji, and Izu-Ogasawara contain Zn, Cu, Pb, Ba, Au, and Ag. This metal association is different from that of midoceanic ridge deposits in which Cu and Zn are rich but Pb, Ba, Au, and Ag are not. These features of back arc basin deposits are similar to those of Kuroko deposits, which are polymetallic stratabound sulfide–sulfate deposits and

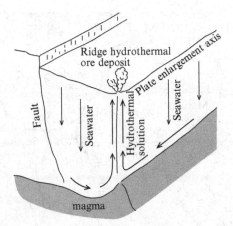

Fig. 4.6 Hydrothermal solution circulating system accompanied by the midoceanic ridge hydrothermal ore deposits

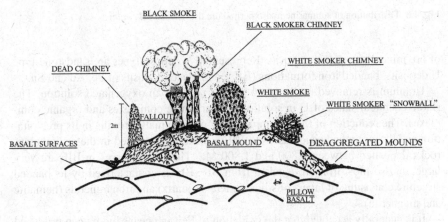

Fig. 4.7 A composite sketch illustrating the variety of structures observed at the different RISE vent sites and the mineral distributions associated with these structures (Haymon and Kastner 1981)

were formed at mid Miocene on the seafloor of the Sea of Japan. Kuroko-type deposits have been formed near the subduction zones (back arc basin, island arc) throughout earth's history since Archean to present time. These deposits are accompanied by various base metals and precious metals such as Cu, Pb, Zn, Fe, Au, Ag, Ga, In, Mo, Sb, As, etc.

Sedimentary Ore Deposits

Sedimentary ore deposits form by precipitation of ore minerals from low temperature seawater, lakewater, or riverwater and by the transportation and accumulation

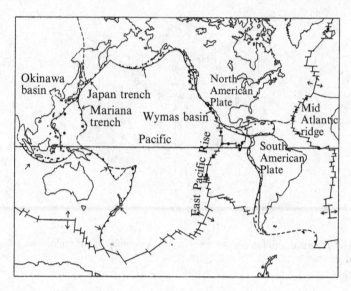

Fig. 4.8 Distribution of submarine hydrothermal ore deposits (*solid circle*)

of ore minerals as placer deposits. Representative deposit types are sandstone-type U deposits, banded iron formations (BIF), and placer deposits of Fe, Au and Sn.

Uranium is removed easily as +6 cations (U^{6+}) in an oxygenated solution. The stable chemical states of U in solution are carbonate complexes and organic complexes. The reduction of these complexes by organic matter results in its precipitation as U^{4+} compounds such as UO_2. BIF is widely distributed in the sedimentary rocks deposited between 3,800 and 1,800 Ma. The iron reserves in BIF are very large, and so most iron ore is taken from BIF. BIF is characterized by its banded structure consisting of layers of silica minerals (quartz) and iron minerals (hematite and magnetite).

It is generally accepted that the oxidation of Fe^{2+} in oceans by oxygen produced by photosynthetic bacteria cause the precipitation of iron minerals on the seafloor, resulting in the formation of BIF at that time.

Large amounts of manganese (Mn) in the form of nodules and cobalt (Co) crust are distributed on the seafloor. Manganese nodules are distributed on deep sea sediments. Cobalt crust occurs on seamounts at shallow depths. In addition to Fe and Mn, base metal elements such as Cu, Ni, Co, and Zn are concentrated in manganese nodules and cobalt crusts (Tables 4.2 and 4.3). They have high potential as mineral resources in the oceans.

Figure 4.9 shows a schematic distribution of mineral resources on the ocean floor, island arcs, back arcs, and continents with respect to topography and geology. As shown in the figure, different types of ore deposits form in different tectonic settings, which are caused by uneven heat flow and plate motions. Besides internal energy, ocean currents caused by external energy from the sun play an important role in the formation of sedimentary deposits.

Table 4.2 Chemical composition of manganese nodule (wt%) (Yoshimatsu and Ogawa 1986)

Element	Indian Ocean West	Indian Ocean East	Mid Atlantic Ocean	Pacific Ocean Baha California seamount	Pacific Ocean California near continent	Pacific Ocean Central seamount	Pacific Ocean Northeast	Pacific Ocean South
Mn	13.56	15.83	16.3	15.85	33.98	13.96	22.33	16.61
Fe	15.75	11.31	17.5	12.22	1.62	13.10	9.44	13.92
Ni	0.32	0.51	0.42	0.35	0.10	0.39	1.08	0.43
Co	0.36	0.15	0.31	0.51	0.01	1.13	0.19	0.60
Cu	0.10	0.33	0.20	0.08	0.07	0.06	0.63	0.19
Pb	0.06	0.03	0.10	0.09	0.01	0.17	0.03	0.07
Ba	0.15	0.16	0.17	0.31	0.14	0.27	0.38	0.23
Ti	0.82	0.58	0.8	0.49	0.06	0.77	0.43	1.01
Depth (m)	3,793	5,046		1,146	3,003	1,757	4,537	3,539

Table 4.3 Chemical composition of cobalt crust (wt%) (Yoshimatsu and Ogawa 1986)

Depth (m)	Mn	Fe	Co	Ni	Cu	Mn/Fe
4,400~4,000	19.7	16.7	0.67	0.24	0.10	1.17
4,000~3,000	25.5	18.0	0.63	0.35	0.13	1.41
3,000~2,400	20.5	19.5	0.69	0.18	0.09	1.05
2,400~1,900	25.5	16.1	0.88	0.41	0.07	1.58
1,900~1,500	24.7	15.3	0.90	0.42	0.06	1.61
1,500~1,100	28.4	14.3	1.18	0.50	0.03	1.90

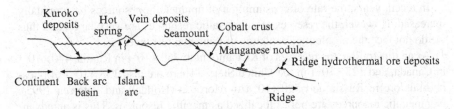

Fig. 4.9 Schematic diagram showing distribution of ore deposits on ocean floor and land (Shikazono 1988)

Weathering Ore Deposits

If rocks are in contact with the atmosphere and rainwater, riverwater, or other surface waters, elements are released from the rocks and secondary minerals such as clay minerals form. The chemical weathering reactions are caused by the presence of H_2CO_3. The reaction between organic acids and rocks and minerals occurs via biological activity. As chemical and biological weathering proceeds, insoluble elements (e.g. Al, Ti) are concentrated in the weathering rocks and soils. Bauxite, an Al ore, forms through the weathering process. It forms in tropical forest regions where there are high temperatures and a great deal of rainfall. In such environments, decomposition of organic matter causes a decrease in the pH of surface waters and creates a reducing environment. In such environments, Fe is released from the rocks, but Al is not, resulting in the enrichment of Al. Al is soluble and Fe is insoluble in oxidizing environments and in lower pH conditions where there is little rainfall. In those environments, laterite containing high amounts of Fe forms. REE are enriched into clay minerals in weathered granitic rocks and soils as iron-adsorption fraction. This type of REE deposits is called ion-adsorption type REE deposits from which more than 90% of ore production of REE (particularly heavy rare earth elements (e.g. Nd, Sm, Dy)) are produced from Southern China.

4.3.4.3 Nonmetallic Mineral Resources

It is difficult to classify nonmetallic mineral resources based on their crustal abundance and genesis. Generally, classification is based on their use. The main nonmetallic mineral resources include fertilizer, which requires N, K, P, and S; chemical products using NaCl and borate minerals; building materials including stone, sand, and gravel; processed rocks that produce cement, lime, gypsum, clays, and glass; and functional materials including nonconductors, polishing materials, and powders. Some nonmetallic mineral resources are taken from nonmetallic ore deposits, but some are not. For example, clays, gypsum, sulfur, and silica are concentrated in nonmetallic ore deposits. Most nonmetallic ore deposits are formed at low temperatures, while some including clays, gypsum, sulfur and silica deposits are of hydrothermal origin.

In recent years, the rate of consumption of nonmetallic resources has gradually increased. However, the reserves of nonmetallic resources are very large, and thus we do not face the problem of depletion of nonmetal resources.

As an illustration, the reserves of KCl amount to 360×10^8 t in Russia, 440×10^8 t in Canada, and 123×10^8 t in the United States. There are also 133×10^8 t of phosphorous ore in Russia and 590×10^8 t in Morocco (Holland and Petersen 1995). Nonmetallic resources are not as localized as metallic resources. This is an advantage when using nonmetallic resources in various ways. New products made from nonmetallic resources instead of metallic resources are being developed.

4.3.5 Water Resources

Water resources are classified according to their usage. They are used for agriculture, city water, industry, power generation, and so on.

Surface water, generally in the form of riverwater or lakewater, and ground water are used extensively but seawater is not much used. Seawater is used for cooling in power generation, in agriculture by changing seawater to fresh water through artificial desalination, and in power generation by using the temperature difference between seawater at the surface and deep seawater. The amount of fresh (terrestrial) water we use has increased very rapidly in recent years, and we may face the problem that water resources could dry up with the depletion of water resources we can use. On the other hand, we are currently developing and using artificial water such as alkali water, ultrapure water, supercritical water, and mineral water.

In the recent past, we have used large amounts of ground water for industry and other uses, resulting in the depletion of ground water reserves. Thus, we have turned to using more surface water instead of ground water. The construction of dams and reservoirs may result in impacts on the ecosystem. Generally, ground water is better for drinking than is surface water.

Rainwater penetrating underground reacts with soils and rocks. During this process, some elements dissolve into the penetrating water. The water, now containing mineral components such as Ca and Mg is good for drinking and other purposes in industry etc. It is important to understand the chemical process and behavior mechanisms of the rainwater penetrating downward and of ground water flow in order to prevent the depletion of high-quality ground water reserves. The hydrologic cycle of the rainwater–ground water–soil–rock system and the influence of anthropogenic activity on it should be understood for this purpose.

Ground water is not a nonrenewable resource like metallic minerals and fossil fuels, but we should take into account that it may become a nonrenewable resource if we overpump large amounts of ground water.

4.3.6 Energy Resources

Energy resources are divided into (1) external energy resources and (2) internal energy resources.

4.3.6.1 External Energy Resources

Fossil Fuels

The main source of energy we use today comes from the fossil fuels oil, coal, natural gas, gas hydrate, oil shale and shale gas. Solar energy is stored in these fossil fuels. The processes through which fossils fuels formed are considered briefly below.

Coal

Land plants are the source of coal. The main components of land plants are cellulose ($C_6H_{10}O_5$) and lignin ($C_{12}H_{18}O_9$). Lignin is the main source of coal. Soon after plants are buried, they change to peat through biological activity. As peat and coal become buried more deeply, their carbon content increases while their volatile (mainly H_2O) content decreases as temperature and pressure increase with depth. Increases in the base metal contents of coal during burial is caused by the reduction of sulfate ions to hydrogen sulfide and the formation of sulfide minerals.

Oil

The source of oil is not clear because no recognizable traces of the source materials are preserved in the oil. However, it is certain that oil formed from organic matter in a reducing (anoxic) subseafloor environment. The reducing condition is necessary to form oil because in oxidizing environments the organic matter oxidizes to yield CO_2 and H_2O. The organic matter that became oil is considered to be biogenic in origin, probably from plankton. Soon after the burial of organic matter-bearing sediments, aerobic bacterial activity enhances the decomposition of the corpses of marine organisms. Then, as temperature and pressure increase, kerogen, which is a highly polymerized amorphous substance of biogenic origin dispersed in sediments, changes to hydrocarbons such as paraffin, naften, and aromatic compounds. As mentioned above, chemical, physical, and geologic conditions favorable to the formation and accumulation of oil are necessary. The occurrence of cap rock and highly porous reservoir rock and high sedimentation rates that protect the decomposing organic matter are important geologic conditions.

Natural Gas

The source of natural gas is similar to that of coal. Coal seams, especially those buried too deep to mine, are both source rocks for natural gas and reservoir rocks where it can be trapped. Natural gas is mainly composed of methane (CH_4).The burning of natural gas releases less CO_2 per unit of energy than the combustion of coal or oil. Oil and coal contain more abundant pollutants including C, N, and S. Natural gas resources are comparable to those of crude oil. Therefore, recently the use of natural gas has increased rapidly.

Gas Hydrate

Gas hydrate is the solid phase of a natural gas (methane)-water compound. It is stable under low-temperature/high-pressure conditions. It has been discovered from under the Atlantic, to the tundra in Siberia and on continental shelves. Near the Japanese Islands it has been found in the Nankai Trough, south of Shikoku and the Kii Peninsula.Gas hydrate reserves are very large, and so gas hydrate has great potential as an energy resource. However, gas hydrates are dispersed in sediments and do not occur as massive bodies, so that mining them is not efficient. During the mining process, the gas hydrate may change to methane gas due to temperature and pressure changes. If methane gas is released to the atmosphere, it may cause global warming.

Other Energy Sources

The force driving wind and ocean currents is solar energy. The kinetic energy of wind and ocean currents can be transformed to electric energy. Solar light energy is stored in solar batteries. Amorphous silicon and polycrystals are used for solar batteries. Solar energy is clean and virtually inexhaustible, and thus its use is expected to increase. However, there are problems with solar energy: (1) its energy density is very low, (2) the efficiency of transforming solar energy is low, (3) its energy varies widely with time, and (4) solar energy is difficult to transport and concentrate. However, it is highly expected to use much renewable energy (solar, wind, geothermal energies etc.) because they are clean, do not produce pollutant, and the depletion does not occur.

Hydrogen Energy

Hydrogen and oxygen react to form water, but the reaction can generate electric power as well. The energy generated by the reaction, $H_2 + 1/2O_2 \rightarrow H_2O$ is used in fuel cells. Water formed by this reaction is clean, and so research in several advanced countries, including Japan, has been focusing on developing technologies to use fuel cells. However, hydrogen is usually produced from organic matter. The reaction also forms CO that is toxic and CO_2 that may cause global warming. The benefits of hydrogen energy are hydrogen energy reserves are very large and there are also vast sources of hydrogen, water in particular. Its demerits are that (1) large amounts of energy are necessary to decompose water to generate hydrogen, (2) transportation and storage of hydrogen energy are difficult, and (3) if we use organic matter to produce hydrogen, CO_2 are also produced, which may cause global warming.

Biomass

Biomass energy comes from organic matter. The problems that limit its use are the low efficiency of transforming its energy and the small amount biomass reserves that are available. It is difficult to get enough wood and other organic matter. Biomass is generally thought to be clean energy because the CO_2 emitted by the burning of biomass returns to plants, which means that biomass energy is clean.

4.3.6.2 Internal Energy Resources

Geothermal Energy

Heat and mass transfer are occurring in the hydrothermal system. The system is composed of heat sources, rocks, vapor, hydrothermal solutions, and passages for the hydrothermal solutions and vapor including fissures, fractures, and pores in rocks.

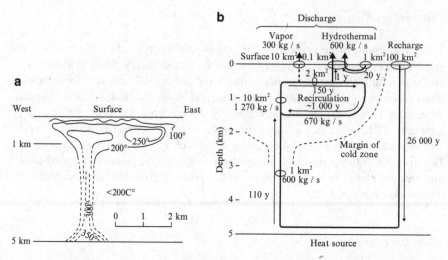

Fig. 4.10 Hydrothermal system (Elder 1966). (**a**) Temperature distribution in Wairakei geothermal system (New Zealand) and (**b**) interpretation of hydrothermal solution movement based on pipe model

Geothermal gradients in hydrothermal systems are very steep compared with in other areas. Geothermal energy is obtained by changing the kinetic and thermal energy of vapor and hydrothermal solutions to electric energy.

An example of a hydrothermal system is shown in Fig. 4.10. As shown in the figure, convection of a hydrothermal solution is occurring. Most hydrothermal solutions originate as meteoric water, but it is likely that hydrothermal solutions originating from magma (magmatic water) and volcanic gas is incorporated into a hydrothermal system. The flow pattern of the hydrothermal solution and the mass transfer in the hydrothermal system (Fig. 4.8) can be estimated by solving the mass conservation law, conservation of momentum law, physical state equations, chemical equilibrium, and kinetic equations. Geothermal energy is generally clean, although sometimes toxic metals such as As are associated with hydrothermal systems. Geothermal energy is used in several countries where igneous activity is intense such as Ireland, New Zealand, the western part of the United States, Mexico, the Philippines, and Japan.

Nuclear Fission Energy

Atoms consist of nuclei and electrons. Nuclei consist of protons and neutrons. The number of protons determines the positive charge of the nucleus and the number of orbiting electrons. The mass of the electrons is negligible compared with that of the protons and neutrons. The total mass of an atom depends on the total numbers of protons and neutrons.

Nuclear fission can occur without neutron bombardment, as radioactive decay. This type of fission (spontaneous fission) is found in a few heavy isotopes. Several heavy elements such as U, Th, and Pu undergo both spontaneous fission (radioactive decay) and induced fission (a form of nuclear reaction).

The nuclear fission has been used in atomic bombs and in nuclear power plants. In nuclear power plant, a neutron reacts with a U^{235} produce U^{236}. This fissions almost immediately into two nuclei and liberates about three neutrons. Not all breakups of U^{235} nuclei produce barium and kryptron. The reaction of a neutron with a U^{235} nucleus is given by

$$_0n^1 + _{92}U^{235} \rightarrow _{92}U^{236} \rightarrow _{36}Ba^{141} + _{36}Kr^{92}\ 3_0n^1 + \gamma \text{ rays + neutrinos}$$

A great amount of electric energy can be produced by such nuclear fission reactions in various types nuclear reactions (BWR (boiling water reactor), PWR (pressurized water reactor), HWR (heavy water reactor), HTGR (high temperature, gas cooled reactor).

Nuclear Fusion Energy

Large amounts of energy are emitted when a light nucleus combines with others and produce a heavy nucleus. However, it is difficult to cause nuclear fusion reactions in the earth's surface environment because the energy level needed to cause the reaction is very high.

Nuclear fusion occurs in the sun and in other stars. The main fusion reactions are the proton–proton (P–P) reaction and the carbon–nitrogen–oxygen (C–N–O) reaction. P–P reactions are occurring in the core of sun, whereas C–N–O reactions occur inside heavier stars. The P–P and C–N–O reactions are given as follows.

P–P reaction:

$$P + P \rightarrow {}^2H + e^+ + \nu$$

$$^2H + P \rightarrow {}^3He + \gamma$$

$$^3He + {}^3He \rightarrow {}^4He + 2P$$

where P is a proton, e^+ is a positron, ν is a neutrino and γ is a gamma ray.

C–N–O reaction:

$$^{12}C + P \rightarrow {}^{13}N + \gamma$$

$$^{13}N \rightarrow {}^{13}C + e^+ + \nu$$

$$^{13}C + P \rightarrow {}^{14}N + \gamma$$

$$^{15}O \rightarrow {}^{15}N + e^+ + \nu$$

$$^{15}N + P \rightarrow {}^{12}C + {}^4He$$

In the P–P reaction, four protons combine to form one He (^{4}He). A quantity of 4.0316 g of protons results in 4.0026 g of He. The mass difference, 0.029 g, corresponds to 2.61×10^{12} joules of energy release.

4.4   Earth's Environmental Problems

After World War II, industrial production increased greatly thanks to developments in scientific research and technology. Accompanying that development, we have faced serious resources and environmental problems. We have recognized that the earth's resources are limited. Various global environmental problems have greatly influenced humans and the ecosystem. In 1962, Rachel Carson (1907–1964) published "Silent Spring (Carson, 1962)", in which she stressed how toxic organic matter destroys ecosystems. Even after that, however, pollution increased in advanced countries. For example, in Japan, there were serious pollution problems such as the pollution of riverwater and soils around the Ashio copper mine area, Minamata disease, which was caused by organic mercury compounds derived from industrial drainage and Itaitai disease, caused by Cd from Pb and Zn mining.

Recent recognition of resources and environmental problems has been based on "The Limit to Growth" commissioned by the Club of Rome and published in 1972. It clearly indicated that the depletion of natural nonrenewable resources and the increased pollution were caused by the rapid growth of the population, and so the growth rate has limits. This was the first report to attempt a quantitative analysis and prediction of temporal changes in the amounts of resources that would be consumed and pollutants that would be produced.

The subject of global environmental problems was first taken up at the World Environmental Conference in Stockholm in 1972. In that Conference, the damage to fish in lakes in Scandinavia by acid rain was discussed. In 1985, the ozone hole was discovered in the stratosphere over the Antarctic. In 1985 the Vienna Convention for the Protection of the ozone layer was agreed upon and in 1987 the Montreal Protocol calling for freon emissions to be reduced by half until 1998 was opened for signature.

In 1988, the global warming problem was discussed by the Summit in Toronto and then James E. Hansen, a famous meteorologist, reported that global warming was caused by an increase in atmospheric CO_2 concentrations. After that point, global warming was widely recognized as one of our most serious global environmental problems. It is obvious that three problems noted above (acid rain, the ozone hole, and global warming) are urgent and should be solved by the development of technology. In addition, there are other global environmental problems such as the disappearance of tropical forests, extinction of species, desertification, and ocean pollution.

In 1992, the Earth Summit (UNCED: UN Conference on Environment and Development was held in Rio de Janeiro, Brazil. In this conference, the Rio Declaration on the Environment and Development, United Nations Framework Convention on Climate Change (UNFCCC, FCCC), and Convention on Biological

Diversity (CBD) were adopted. In 1997, the Kyoto Protocol was made with regard to reduction of greenhouse gases such as CO_2. According to the protocol, Japan's emission of CO_2 must be reduced by 6% of the 1990 emission level by 2008–2012.

4.4.1 Global Warming

Global warming is thought to be caused by increased concentrations of greenhouse gases such as CO_2 in the atmosphere.

Sunlight contains visible light with wavelengths in the range of 0.17–4.0 μm. The energy in sunlight is not absorbed by the N_2, O_2, Ar, H_2O, or CO_2 molecules in the atmosphere and so reaches the earth's surface. Sunlight energy is, however, absorbed and reflected by aerosols and clouds. Infrared rays with long wave lengths radiate from the earth's surface.

Greenhouse gases such as CO_2 and H_2O absorb the infrared radiation from the surface and emit part of it back as thermal energy, resulting in an atmospheric temperature increase near the earth's surface. This is called the greenhouse effect.

4.4.1.1 Increased Atmospheric CO_2

Increased atmospheric CO_2 caused by human activity is mainly due to industrial CO_2 emissions and deforestation. Enlargement of farms, the occurrence of forest fires, and the use of slash-and-burn agriculture in developing countries also contribute to this increase. Most of CO_2 emissions from industrial activity come from the burning of fossil fuels. CO_2 emissions from industrial activity have been increasing exponentially during the last 50 years. The recent warming trend correlates well with the increase of CO_2 from emissions since the industrial revolution of the nineteenth century. Figure 4.11 shows the change in atmospheric CO_2 concentration measured at the top of Mauna Loa (a 4,300 m high volcano in Hawaii) observatory since 1958. As shown in this figure, the atmospheric CO_2 concentration has increased from 315 ppm in 1957 up to 380 ppm in 2008. Similar upward CO_2 trends have been measured at many different stations around the globe. The average atmospheric CO_2 concentration before the industrial revolution of the nineteenth century was about 280 ppm, shown by analysis of CO_2 in ice cores from Greenland and Antarctica. Seasonal variation in atmospheric CO_2 concentrations can also be observed. Atmospheric CO_2 in spring and summer is higher, and lower in the autumn and winter. In spring and summer, atmospheric CO_2 is removed by the photosynthetic activity of plants and CO_2 is emitted by the decomposition of dead plants in autumn and winter.

If all the CO_2 emitted by human activities were stored in the atmosphere, we could estimate the annual rate of atmospheric CO_2 concentration increase because the annual CO_2 emissions and total amount of atmospheric CO_2 can be estimated. The measured atmospheric CO_2 concentration is about one half of the estimated

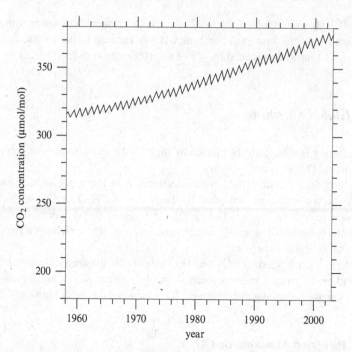

Fig. 4.11   Atmospheric CO_2 concentration in recent years based on direct determinations at Mauna Loa Observatory. Data from Keeling and Whorf (2003) (Marini 2007)

concentration. This means half of the CO_2 added annually to the atmosphere is removed within a short time. There are two processes that could achieve this reduction. One is dissolution of atmospheric CO_2 into the oceans. The other is removal of atmospheric CO_2 by photosynthesis in forests plants. The forests in the northern hemisphere forest probably reduce atmospheric CO_2 considerably and may be the major sink for the "missing CO_2".

4.4.1.2   Other Greenhouse Gases

Gases besides CO_2 that can promote the greenhouse effect include freon (the common name for a group of chlorofluorocarbon compounds including $CFCl_3$), halons (CF_3Br etc.), methane (CH_4), water vapor (H_2O), ozone (O_3), and nitrous oxide (N_2O). These gases absorb infrared radiation but do not emit radiant energy. The contribution of each greenhouse gas to global warming is different: carbon dioxide is responsible for 49%, methane 18%, nitrous oxide 6%, freon 14%, and the others contribute 13%. Chlorofluorocarbons (F_{11} ($CHCl_3$) and F_{12} (CF_3Cl_2)) are 100% anthropogenic, but other gasses are both natural and anthropogenic in origin.

Fig. 4.12 Prediction of worldwide average (atmospheric temperature until 2060 (Hansen et al. 1988))

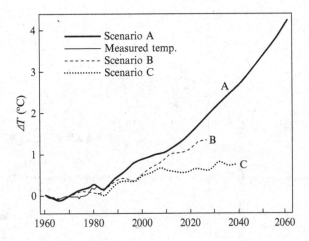

4.4.1.3   Prediction of Temperature Increases in the Future due to Greenhouse Gases

There have been several predictions of atmospheric temperatures that take into account the effect of greenhouse gases. Figure 4.12 shows one example, the prediction by Hansen et al. (1988), which predicts the smallest possible temperature increase during 2000–2033 as 0.3°C, meaning a rate of increase of 0.9–1.0°C per 100 years. This rate is considerably higher than what has occurred in the past 100 years (0.4–0.6°C). However, the prediction of atmospheric temperature is very difficult due to the large uncertainty in the estimation parameters. If we knew how much atmospheric CO_2 would increase, we could estimate the increase in temperature because the greenhouse effect of CO_2 is well studied. However, if temperature increases, the rate of evaporation of seawater increases, resulting in a larger greenhouse effect due to an increase in absorption of infrared radiation by water vapor in the atmosphere. Also, the area of the polar ice sheets decreases with increasing temperature. Due to this decrease, the albedo of the earth's surface (the fraction of the sun's energy that is reflected back into space) decreases because the albedo of ice sheets is higher than other surfaces such as rocks, forests, and oceans. The melting of methane hydrate in tundra that occurs as temperatures increase also generates methane gas emissions, promoting the greenhouse effect. These examples all exhibit positive feedback. Cloud cover may be involved in both positive and negative feedback cycles. It absorbs infrared radiation and heat in the atmosphere but also reflects the sun's energy. Its role as a factor controlling atmospheric temperature is not well understood.

As mentioned above, it is certain that we have to take into account various positive and negative feedback systems accompanying the change in atmospheric CO_2 concentration when we predict climate change in the future.

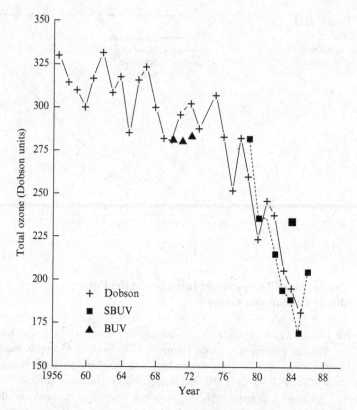

Fig. 4.13 October monthly mean total ozone measurements over Harley Bay Plusses are Dobson measurements by Farman et al. (1985). *Triangles* are Ninbus-4 BUU measurements from 1970 to 1972; *squares* are Nimbus-7 measurements from 1979 to 1986 (Stolarski 1988; Holland and Petersen 1988)

4.4.2   The Destruction of the Ozone Layer

Recently it was discovered that an ozone hole is present over Antarctica. The ozone concentration over Antarctica has gradually decreased since 1956 (Fig. 4.13). Atmospheric temperature increases from 10 km above the ground to about 50 km where the atmospheric temperature is at a maximum. This is caused by the absorption of solar ultraviolet radiation by ozone. Solar ultraviolet radiation is harmful to living matter and thus the ozone layer protects ecosystems and humans. A decrease in the ozone concentration increases the incidence of diseases like skin cancer and cataracts.

In the ozone layer solar ultraviolet radiation dissociates molecular oxygen into two oxygen atoms. The subsequent reaction of single oxygen atoms with molecular oxygen can produce ozone by the reaction

$$O + O_2 + M \rightarrow O_3 + M$$

where M indicates other components involved in this reaction.

The excess oxygen atom is removed by the reaction

$$O + O_3 \rightarrow O_2 + O_2$$

Ozone is unstable compared with molecular oxygen. At an altitude of 30 km, one ozone molecule exists per 3×10^4 oxygen molecules. This is a very low density compared with the equilibrium number. In the ozone layer, ozone is produced by solar ultraviolet radiation, whereas nitrous oxide dissociates it. Nitrous oxide is generated by microorganic activity. The steady state condition with regard to ozone concentration is a balance between ozone production and decomposition.

In recent years, the ozone concentration has decreased due to anthropogenic influences. Freon (chlorofluorocarbons) produced by humans decomposes ozone. Ultraviolet radiation is absorbed by chlorofluorocarbons, which then generate chlorine atoms. Chlorine atoms react with ozone to decompose it to ClO and O_2 as follows.

$$Cl + O_3 \rightarrow ClO + O_2$$

ClO decomposes to Cl and O by reacting with O as

$$ClO + O \rightarrow Cl + O_2$$

Cl reacts with methane and finally becomes HCl, which is removed by rain. It takes several years for chlorofluorocarbons from ground surface to reach the ozone layer. Therefore, it is certain that would take at least several years to decrease the chlorofluorocarbon concentration in the ozone layer after halting their emission.

Freon is stable and accumulates in the atmosphere. It goes to the troposphere and then rises to the stratosphere where it is decomposed by solar ultraviolet radiation.

In addition to freon, nitric oxide decomposes ozone by reacting with ozone to produce nitrogen dioxide and molecular oxygen:

$$NO + O_3 \rightarrow NO_2 + O_2$$

Nitric oxide was contained in the exhaust gas of the supersonic transport.

4.4.3 Acid Rain

Rainwater acidification mechanism:

The term acid rain is used to describe rain having a low pH and sometimes to denote fine acid precipitates.

Rainwater dissolves atmospheric CO_2. The pH of CO_2-bearing rainwater becomes slight acidic (pH $= 5.65$). This acidification of rainwater by natural processes is not called acid rain. According to an investigation of acid precipitates from the USA, acid rain is defined as rainwater with a pH lower than 5.0.

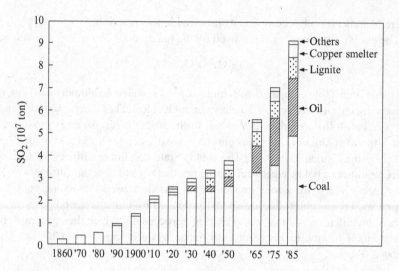

Fig. 4.14 SO$_2$ emission with time (Environmental Agency Earth Environment Work Shop 1990)

The main artificial compounds causing acidification include sulfur oxide and nitrogen oxide. Recently, the anthropogenic emissions of these compounds have been increasing very rapidly (Fig. 4.14). Sulfur oxide gases are dominantly SO$_2$ generated by the oxidation of sulfur in coal and oil during burning.

Sulfuric acid is formed by the following two types of reactions.

$$SO_2 + O_3 \rightarrow SO_3 + O_2,\ SO_3 + H_2O \rightarrow H_2SO_4 \left(\text{sulfuric acid}\right) \rightarrow \left(H_2SO_4\right)_n \left(\text{aerosol}\right)$$

$$SO_2 + 1/2O_2 \rightarrow SO_3,\ SO_3 + H_2O \rightarrow H_2SO_4 \left(\text{sulfuric acid}\right) \rightarrow \left(H_2SO_4\right)_n \left(\text{aerosol}\right)$$

Methods to reduce the emission of sulfur dioxide gas and sulfuric acid are (1) The burning of fossil fuels containing smaller amounts of sulfur, and (2) removal of SO$_2$ from combustion gases.

Nitrous oxide is produced by the oxidation of nitrogen in high temperature air used for combustion. It also comes from exhaust gas from cars and NO$_x$ compounds come from fertilizer in soils.

Damage by Acid Rain:

Lakes, ponds, rivers, forests, agricultural crops, excavations, buildings, and soils are all damaged by acid rain. For example, surface waters like lakes, ponds, and rivers are acidified by acid rain, resulting in fish losses. The reaction of acid rain with soil promotes the dissolution of Al out of the soil. H$^+$, SO$_4^{2-}$, and base metals (Zn, Cd, etc.) in soil water all increase due to acid rain reactions.

The reaction of acid water with soil dissolves cations such as Ca^{2+}, Mg^{2+}, and K$^+$, which are essential for the growth of crops. They leave the soil and enter the soil water. Crops may suffer a lack of phosphorus, sulfur, calcium, magnesium, and molybdenum as a result.

The damage done by acid rain is widely recognized in Europe, the USA, and Canada but not in Japan, although acid rain is widespread in Japan. The reasons for very little damage by acid rain in Japan are (1) the climate in Japan is characterized by large amounts of rainfall and high humidity, (2) the pH of rainwater penetrating underground increases very rapidly as it reacts with the soil and volcanic ash, tuff and other rocks that react quickly with aqueous solutions to buffer the pH, (3) rainwater penetrates easily underground because of Japan's thick soils, and (4) acid mists with pH levels lower than acid rain's are not common.

There are various sources of riverwater and lakewater; ground water, rainwater, seepage water, and hot springs. The proportion of these source waters in rivers and lakes depends on geology, geography, and climate. In Japan, the dominant rocks are young volcanic rocks. They are highly permeable due to fissures, fractures, faults, and their high porosity. Intense weathering of these rocks, Japan's relatively warm climate and high biological activity combine to produce thick soils. The soils are highly porous, resulting in efficient penetration of rainwater underground. The penetrating rainwater's pH and chemical composition change, and it becomes ground water. Shallow ground water migrates from higher elevations to lower ones and then seeps from the surface to become riverwater. Thus, the proportion of ground water in riverwater is relatively high in Japanese riverwater. That means the pH of Japanese riverwater is not low, being buffered by water–rock interactions underground.

In contrast, in continental regions (the United States, Canada, and Europe) the dominant rocks are granitic and metamorphic. They react much less with water. Thus the pH and chemical composition of the water in continental regions change much less than in regions where the rocks and soils are of volcanic origin such as Japan. The permeability and porosity of these rocks are low that rainwater does not tend to penetrate deep underground. Rainwater penetrates through very shallow parts such as regoliths, weathered rocks, and thin soils and flow within them. On the ground surface, acid substances such as H_2SO_4 and aerosols accumulate. Therefore, if rain falls onto the surface, the rainwater becomes more acidic. The main sources of riverwater in continental regions are acid rainwater and acidic shallow surface water. The proportion of deep ground water with higher pH is generally small. However, in continental limestone regions the pH of rainwater, riverwater, and lakewater is high because of the rapid dissolution of carbonates ($CaCO_3 + H^+ \rightarrow Ca^{2+} + HCO_3^-$) in the limestone.

4.4.4 Soil Problems

Soil problems include (1) surface soil loss, (2) accumulation of salts on the surface of the soil, (3) release of elements from soils, and (4) soil contamination.

Surface soil loss is enhanced by anthropogenic activities such as deforestation, mining, agriculture, and acid rain. Accumulation of salts on the surface of soils is related to desertification. The main causes of this are thought to be over-pasturage, irrigation, deforestation associated with irrigation, construction of dams, enlargement of urban areas, etc. Climate change may also cause this problem. Desertification intensifies the evaporation of lakewater and the precipitation of salts onto dry lake bottoms. These salts are dispersed by the wind to the surroundings and precipitate onto the ground surface. Agricultural crops do not tend to grow in saline and alkaline soils. Common salts that accumulate near the surface of the soil are $CaCO_3$, Na_2CO_3, $NaCl$, KCl, and $CaSO_4 \cdot 2H_2O$.

If carbonates dissolve into water, the following reaction occurs.

$$CaCO_3 + H^+ \rightarrow Ca^{2+} + HCO_3^- + CO_2$$

This reaction results in a lower concentration of H^+, or higher pH.

Fertilizers, insecticides and toxic base metals from waste contaminate soils. Part of these soil contaminants is released but the rest remains in the soil. Various substances such as toxic organic matters and base metals can adsorb onto soil very readily.

Soils contain a mixture of organic matter, and primary and secondary minerals like clay minerals, which all have different adsorption capacities. The combination of materials makes it hard to remove contaminants from the soils. For example, iron (oxy) hydroxides ($FeOOH$, $Fe(OH)_3$) in soil tend to adsorb base metal ions. Most agricultural chemicals tend to undergo hydration, combining with water molecules, and adsorb onto organic matter. Thus, these chemicals accumulate in organic matter-rich soils. The mechanism by which chemicals adsorb onto organic matter is complex, but it is certain that organic matter plays an important role in adsorption.

Oxidation-reduction conditions are also an important factor controlling the behavior of chemicals and base metals in soils. Oxidation-reduction conditions depend on the amount and kinds of organic matter present and the rock type. In reducing environments base metals are fixed as sulfides, while in oxidizing environments iron (oxy) hydroxides form and base metals are adsorbed onto them. Base metals are contained in industrial waste, agricultural chemicals, waste, and drainage from mines. Cd, Cu, and As in agricultural chemicals and wastewaters are identified as toxic elements and their safe levels are determined by environmental law.

Reducing ground water is formed by reaction with anthropogenic organic matter. Such ground water dissolves iron and manganese from the rocks and releases arsenic adsorbed onto the iron hydroxides. High pH levels also cause arsenic to be released from iron (oxy) hydroxides.

Mercury contamination is the most serious problem among the base metal contaminations. The toxicity of mercury depends on the chemical compound containing it. Liquid mercury is nonpoisonous, but its vapor is deadly. Organometallic mercury is also poisonous. Methyl mercury (CH_3Hg^+) and dimethyl mercury (($CH_3)_2Hg$) in fish is very poisonous and causes severe mercury poisoning.

Drainage from paper mills and industries producing sodium hydroxide contains mercury, which is transported by rivers and accumulates in lake bottoms and in ocean bottom sediments.

The mercury in sediments changes to methyl mercury, which can dissolve in riverwater, lakewater and seawater. The mercury accumulates in fish and other living marine matter. If the mercury is fixed as mercury sulfide like cinnabar or HgS in a reducing environment where it does not dissolve, it is not decomposed by bacterial activity and thus is not poisonous.

The transportation of base metals by riverwater and ground water is the main cause for the base metal contamination of soils and sediments. Aerosol fall also accounts for part of the high concentration of base metals (Cd, Cu, Pb, and Zn) on the ground surface.

Plants concentrate the base metals and rainfall and subsequent surface water flow removes and disperses them. Transportation and enrichment of the base metal elements occur near the soil surface.

Dioxin and other toxic organic substances are accumulated in living marine living matter near the coast.

4.4.5   Water Pollution

Surface waters are polluted by various wastes. Polluted wastewaters from various sources such as industries and mines enter rivers. Riverwater polluted by toxic base metals, organic substances, nitrogen, phosphorus, and other pollutants enters the ground water, lakes, and oceans. Damage like eutrophication and red tides in lakes and coastal oceans are caused by nitrogen and phosphorus pollution. Ground water is contaminated by toxic base metals like Cr^{6+} and toxic organic matter such as tetrachloroethylene, dioxine, and trichloroethylene from industrial sources.

Ocean contamination is caused by inputs of contaminated riverwater, oil from tankers, dissolution of toxic compounds such as ship paints containing organotin compounds, and disposal of various wastes. Anthropogenic toxic substances accumulate in the closed oceans, resulting in their enrichment in living marine matters and damage to human health.

4.4.6   Waste Problems

Humans take various natural resources from natural systems, use them, and emit various wastes back to nature. Humans have been developing technologies to explore for natural resources like ore deposits, learning about the genesis of natural resources like ores and oil, and depleting those natural resources.

In contrast to source problems, the sink, or waste, problems have not been well researched. The important point in sink problems is how to reduce the amount of waste. For example, considerable effort has been put forth recently to reduce greenhouse gas emissions, particularly since the Kyoto Protocol in 2005. However, it is impossible to consume and recycle all resources in the system of human society. Therefore, we need to consider how to dispose of waste to the natural environment system.

4.4.6.1 Treatment and Disposal of Industrial and General Wastes

Industrial wastes, such as waste oil, cinders, waste plastic, and paper, and general wastes including garbage and bulky refuse differ by specific wastes.

Industrial waste problems are the effects of surface and underground disposal. Wastes disposed underground dissolve into the ground water. The contaminated ground water migrates and pollutes soils and living matter over a long time. We need to develop technologies that prevent the dispersion of wastes from their disposal sites.

We face the problems of gathering, transporting, treating and disposing of general wastes. Disposal site selection, illegal disposal, migration of toxics from landfills, and cross-border movement of toxic wastes are all problematic.

4.4.6.2 Disposal of Nuclear Waste

We also face the problem of how to dispose high-level nuclear waste derived from nuclear power plants. If we put these nuclear wastes in shallow underground sites or in surface environments, radionuclides can migrate and influence humans and eco-systems for a long time (10^3–10^6 years). Therefore, we need to dispose of them deep underground (e.g. more than 300 m deep in Japan).

Wastes disposed underground interact with the surrounding environment. Over the long-term, for example, ground water dissolves wastes and their radionuclides can migrate long distances. We need to predict how far radionuclides will migrate and how radioactive they will after some number of years. The radioactivity at the surface should be less than set safety levels. It is important to study the migration and retardation mechanism of radionuclides in geologic environment (ground water, host rocks, soils). The long-term process of radionuclide migration is influenced by various geologic events (e.g. earthquake, volcanic activity, faulting, uplift, erosion, climate change, seawater level change). We estimate that it takes more than a thousand years for radionuclides to migrate from a waste disposal site to the surface. It is generally believed that long-term isolation of radionuclides from humans and the biosphere is possible if waste is disposed in stable deep underground.

4.4.6.3 CO_2 Disposal

The main cause of global warming is thought to be anthropogenic CO_2 emission, and so this must be reduced. Efficient CO_2 utilization and energy transformation, and utilization of alternative energy sources such as sunlight are useful in reducing CO_2 emissions. CO_2 is fixed naturally via photosynthesis in terrestrial plants and marine organisms and by dissolution into the ocean. Disposal of CO_2 in deep seawater and underground CO_2 sequestration in ground water aquifers, oil reservoirs and underground coal mines are also useful methods of handling CO_2. Zero emission of anthropogenic CO_2 is impossible, and so some CO_2 will always be emitted to the natural environment. Therefore, scientific understanding of the behavior of CO_2 in earth's environment is essential in solving the CO_2 and global warming problems. CO_2 injected to deep (ca. 1,000 m) underground aquifer dissolves into ground water (solubility trapping) becoming acid solution by

$$CO_2 + H_2O \rightarrow H_2CO_3 \rightarrow H^+ + HCO_3^-$$

The acid CO_2 bearing ground water dissolves Ca^{2+}, Mg^{2+} and Fe^{2+} by

$$CaO(MgO, FeO) + 2H \rightarrow Ca^{2+}(Mg^{2+}, Fe^{2+}) + H_2O$$

where CaO (MgO, FeO) is CaO (MgO, FeO) component in minerals.

With the proceeding of these water–rock interaction, the concentrations of Ca^{2+}, Mg^{2+}, Fe^{2+} and HCO_3^- increase, resulting to the formation of carbonates ($CaCO_3$, $CaMg(CO_3)_2$, $FeCO_3$) (mineral trapping). The long-term variation of amounts of carbon with time fixed by solubility trapping and mineral trapping can be calculated by computer simulation (Shikazono et al. 2009). In addition to scientific and technological research, social procedures such as emission trading, carbon taxation, and improvement of environmental laws must be undertaken.

4.4.7 Provisions for Earth's Resources and Environmental Problems

4.4.7.1 Technological Treatments

Energy Conservation and Efficient Energy Use

If we conserve energy and use it efficiently, the amount of fossil fuels we consume could decrease.

Efficient energy use in industrial activities, improvements in electrical fittings, increases in gas mileage, social system improvements, cogeneration, and the use of new energy sources can contribute to energy conservation and reduce CO_2 emissions.

Much heat is lost in the process of using energy, and so we should seek ways to efficiently use waste heat. Gas emitted from thermal power generation plants is very hot, and it escapes into the atmosphere. Systems that use waste heat are called cogeneration systems. Relatively low-temperature waste heat from incinerators, sewage disposal, and air conditioners has potential for use in cogeneration applications.

Recycling

Recycling of waste decreases waste emission and contributes to preserving resources. However, we cannot recycle every kind of waste. Some metals such as Al and rare metals are already recycled to a considerable extent. However, it is difficult to separate

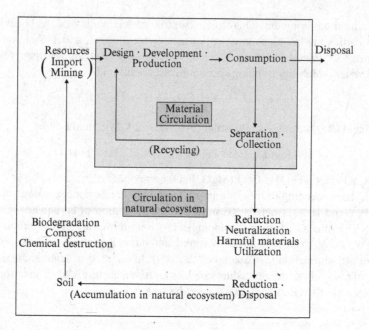

Fig. 4.15 Earth's surface environment and humanity combined system (modified after Takasugi 1993)

and recover metals from alloys in waste streams. The recycling of other wastes (domestic, industrial, and nuclear wastes, and toxic gases) is very difficult. Domestic waste contains various substances and is difficult to separate. Accordingly, large amounts of such waste are burned. Technologies to remove toxic gasses have been developed to prevent contamination. For example, SO_2 gas can be removed as calcium sulfate. CO_2 sequestration technologies such as underground storage can be used to reduce CO_2 gas emission to the atmosphere.

We developed industry by using large amounts of natural resources, which brought increased environmental loads. To overcome resources and environmental problems, soft energy passing as proposed by E.B. Robins (1976) instead of the previous hard energy passing should be used. This aims at improving energy use efficiency of and construction of societies based on natural energy usage.

Recently, it was proposed that the earth's surface environment and humanity should be thought of as a combined system, as shown in Fig. 4.15. In such a system, a large proportion of waste is recycled and the waste emitted to the natural system is reused. For example, we can use soils for agriculture produced by transforming wastes through bacterial activity.

There are wastes that are difficult to recycle and reuse. For example, nuclear waste and CO_2 originating from nonrenewable energy resources cannot be recycled, reclaimed or reused. Therefore, waste from nonrenewable resources flows only one way and does not circulate. Therefore, we need to investigate the behavior of waste, including their mobility and the chemical reactions they undergo in natural systems. Then we may be able to change the natural system using technology and construct a human society system that can coexist, or be in harmony, with the natural system. Technology is very important for this aim, but political, environmental and sociological treatments are also necessary.

4.4.7.2 Political and Economic Treatments

Recently, various economic treatments have been applied to environmental and waste problems. For instance, implementations of surcharge systems, grants, new market systems and the foundation of deposit/refund systems and ecobusinesses may ease environmental and waste problems in different settings.

There has been much discussion about international agreement concerning the reduction of contaminants. Examples include the drafting of international treaties for acid rain (1985 Helsinki Protocol) and freon gas (1985 Vienna Convention and 1987 Montreal Protocol, etc.).

In addition to CO_2, freon, SO_2, and nitrous oxide, maximum emissions and concentration levels of various contaminants need to be determined in the future. Among the various gases, the reduction of CO_2 emissions is the most difficult problem. Although the Kyoto Protocol was drafted in 1997, CO_2 emissions from Japan and other advanced countries have continued to increase.

It is essential to understand that political, economic, and sociological problems should be open to the natural system. The object of early economic disciplines was restricted to market systems. However, since a warning by B. Fuller in 1969 when he described the "Earth Spaceship" concept, economics has changed. For example, A.V. Kneese et al. (1994) applied a material balance approach that considers economics as open to natural systems including the ecosystem. Their approach was based on the idea that the human economic system consists of (1) energy exchange, (2) material processing, (3) consumption of energy and materials, and (4) processing and disposal of wastes. Material and energy are exchanged among these four processes. It is clear by this approach that economic activity is significantly constrained by earth systems including ecosystems (Ueda et al. 1991). This environmental economics focused on humanity, not taking the role of earth's systems, including humanity, as constraints on human activities. In other words the objects of that type of environmental economics were limited to the human society system and part of earth's surface environment, limited to the atmosphere, water, soils, and living things. Earth's system consists not only of these subsystems but also of the geosphere, including the crust, mantle, and core. It is important to understand how the deep and shallow earth environments interact in order to think about earth's resources and waste disposal problems, which are important subjects in environmental economics (Fig. 4.16).

4.4.7.3 Earth Environment and Resource Ethics

We need to consider on what temporal and spatial scales we are responsible for dealing with environmental and resource problems. To answer this question

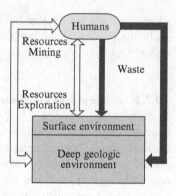

Fig. 4.16 Interaction between humans, surface environment and deep geologic environment

we should understand earth environment ethics, which include the following principles:

1. Not only humans but also nature has a right to survive.
2. Today's generation should not restrict future generations' possibility of surviving.
3. Preservation of earth's ecosystem precedes the other purposes.

This system of ethics has several problems. "Nature" is not clearly defined. "Nature" in this code of ethics seems not to include the solid earth. However, the earth system consists of the solid earth as well as the atmosphere, the biosphere, and the hydrosphere. Principle 2 concerns the environment, which we should preserve for the future. However, people today should also preserve natural resource reserves. Depletion of natural resources has great bearing on the possibility that people will be able to survive in the future. We need to have ethics that deal with resources as well as the environment. We will call such a system "earth environment and resource ethics". Earth environment and resource ethics is based on equality between different generations. There is also the idea that members of the same generation should be equal as well.

There is also the problem of uneven distribution of resources. If the reserves of particular metals and fossil fuel resources are concentrated in a few countries, are those resources owned only by these countries? A few advanced countries use most of the natural resources consumed by the entire world. We do not have international agreements about equality in the use of natural resources or about dealing with environmental problems between different generations, different countries, or different areas.

4.5 Chapter Summary

1. The main nature–human interactions are natural disasters, environmental problems, and natural resource problems.
2. Natural disasters are classified into disasters caused by external forces (the atmosphere, oceans, rivers, etc.) and those caused by internal forces (volcanism, earthquakes, etc.)

3. Resource problems include depletion of resources and environmental problems associated with the exploitation of natural resources.
4. Natural resources are classified into minerals, water, soils, fossil fuels, nuclear energy, sun energy, biomass energy, etc.
5. Natural resources are generally divided into renewable (solar energy, biomass, water, etc.) and nonrenewable resources (metallic and non-metallic minerals, uranium, and fossil fuels). However, in recent years, some renewable resources such as ground water are becoming nonrenewable.
6. Earth's environmental problems are characterized by long-term, global features. These include global warming, atmospheric ozone depletion, the destruction of tropical forests, extinction of living things, and desertification.
7. Currently, we are facing problems concerning various wastes, including industrial emissions, CO_2, and nuclear waste.
8. We have to consider the various aspects of environmental and waste problems, including those of technology, economics, politics, and ethics.
9. Earth's environment and resource ethics is a basis for constructing a sustainable human society and ecosystem.

References

Carson R (1962) Silent spring. Houghton Mifflin, Boston

Elder JW (1966) Heat and mass transfer in the earth: hydrothermal systems. Bulletin (Department of Scientific and Industrial Research, New Zealand), No. 169

Environmental Agency Earth Environment Work Shop (1990) Politics and economics of Earth environment - New globalism and Japan. Diamondosya (in Japanese)

Farman JC, Gardeiner BG, Shankin JD (1985) Large losses of total ozone in Antarctica reveal seasonal ClOx/NOx interaction. Nature 315:207

Fuller B (1969) Operating manual for spaceship Earth. Southern Illinois University Press, Carbondale

Hansen J, Fung I, Uasis A, Rind D, Lebedeff S, Ruecly R, Russell G (1988) Global climate changes as forecast by the Goddard Institute of Space studies three-dimensional model. J Geophys Res 93:9341

Haymon RM, Kastner MC (1981) Hot spring deposits on the East Pacific Rise at 21°N: preliminary description of mineralogy and genesis. Earth Planet Sci Lett 53:363–381

Holland HD, Petersen U (1995) Living dangerously. Princeton University Press, Princeton

Kanamori H (1978) Quantification of earthquakes. Nature 271:411–414

Keeling CD, Whorf TP (2003) Trends: a compendium of data on global change. Carbon Dioxide Information Analysis Center, Oak Ridge National Laboratory, Oak Ridge

Marini L (2007) Geological sequestration of carbon dioxide. Elsevier, New York

Meadows DH, Meadows DL, Renders J (1992) Beyond the limits. Green Publication Company, Post Mills

Nishimura M (1991) Environmental chemistry. Shokabo, Tokyo (in Japanese)

Nishiyama T (1993) Resources economics. Chuokoron-shinsha, Tokyo (in Japanese)

Robbins E (1979) Soft energy path, Jijitsushinsya, (in Japanese) (Murota,Y and Tsuchiya, H trans)

Shikazono N (1988) Chemistry of Kuroko. Shokabo, Tokyo (in Japanese)

Shikazono N, Harada H, Kashiwagi H, Ikeda N (2009) Dissolution of basaltic rocks and its application to underground sequestration of CO2; estimate of mineral trapping by dissolution-precipitation simulation. J Petrol Mineral Econ Geol 38(5):149–160 (in Japanese)

Siever R, Grotzinger J, Jorden TH (2003) Understanding Earth. W. H. Freeman and Company, New York

Skinner BJ (1976) Earth resources, 2nd edn. Prentice Hall, Englewood Cliffs

Skinner BJ, Porter SC (1987) Physical geology. Wiley, New York

Stolarski RS (1988) Changes in ozone over the Antarctic. In: Rowland FS, Isaksen ISA (eds) The changing atmosphere. Wiley, New York, pp 105–119

Takasugi S (1993) Challenge to environmental problem. NHK Books, Tokyo (in Japanese)

Ueda K, Ochiai H, Kitabatake Y, Teranishi T (1991) Environmental economics. Yuhikaku, Tokyo, p 258 (in Japanese)

Yoshimatsu S, Ogawa Y (1986) A worder rare metal and rare Earth. Kodansya, Tokyo (in Japanese)

Chapter 5
The Universe and Solar System

In the previous chapters, earth's subsystems, including humans, were considered. It is well known that the entire earth and its subsystems have changed between the earth's birth and the present. This change (evolution) is caused by the internal and external effects of interactions among earth's subsystems and by energy and materials input from external systems, meaning the solar system and universe. The early earth system in particular was influenced considerably by the universe. In order to consider the origin of the earth and planets in the solar system, the formative processes at work in the universe should be understood.

Keywords Asteroids • Earth-type planets • Evolution of stars • Giant planets • Planet • Protosolar system • Universe

5.1 Origin and Evolution of the Universe

5.1.1 Big Bang Universe

It is generally accepted that our universe is expanding and that it started as a big bang universe. The big bang universe is an initial universe with a comparably small space in a very high density/high temperature state where no galaxies or stars existed. It is thought that the universe expanded very rapidly from this primitive state, with the generation of elements, the earth, sun, stars, and galaxies following.

In 1922, Alexander Friedman solved Einstein's gravitational field equation, leading to the conclusion that the universe is always expanding or shrinking. This model contrasts with Einstein's static universe model. At present, Friedman's model is more widely accepted than Einstein's model from evidence of universal expansion, verified by Hubble (1929). He discovered the red shift in the spectral lines of the galaxy. The redshirt becomes more pronounced as distance increases. This indicates that the galaxies are retreating from the earth and the universe is expanding. If the rate of retreat was constant, the universe occupied a comparatively small space at 160 Ga.

N. Shikazono, *Introduction to Earth and Planetary System Science*,
DOI 10.1007/978-4-431-54058-8_5, © Springer 2012

In 1946, George Gamow proposed the big bang universe theory in which the primary state of the universe was highly dense, hot, and localized at one point. It then suddenly expanded in the big bang at about 160 Ga. Penzias and Wilson (1965) found that our universe is filled with microwave radiation. This finding is strong evidence supporting the big bang universe theory.

Although a steady state universe theory has been proposed, the big bang theory is more widely accepted. According to the $\alpha\beta\gamma$ theory of Gamow and others, all the elements were formed in the big bang. However, later theoretical calculations indicated that only hydrogen and helium were formed then, not the other heavier elements. The H/He ratio at that time was calculated to be 3% by weight, equal to that of the universe today.

Friedman showed the relation between time and temperature during the several minutes after the big bang to be $10^{10}\sqrt{t}$ K. Thus, temperature after 1 min was 10^{10} K. Starting from the big bang, substances were synthesized and by 10^{-4} s after the big bang, the four fundamental forces—the electromagnetic, weak, and strong forces, and gravity—had formed. Rapid expansion (inflation) occurred between $10^{-3.5}$ and $10^{-3.3}$ s after the big bang according to the inflation universe model, and the universe expanded by 10^{51} times.

5.1.2 Evolution of Stars

In light stars nuclear fusion reactions are occurring. Hydrogen changes to helium and then to carbon and oxygen (Sect. 4.3.5). In the cores of heavy stars, Mg, Si, Fe and elements with nuclei lighter than Fe are formed. The explosive burning of supernovae leads to the formation of large amounts of neutrons, which react with Fe in the central part of star to heavier elements (i.e. U).

Heavier elements such as Ag, Au, or Pb are formed in a variety of reactions. In all of these, elements and energies are consumed, leading to the creation of large stars. Some die as supernovae explosions. Others, however, that formed in the initial stages of galaxy formation have kept growing until now. Such stars have small masses and long life times. In general, the life time of heavy stars is short compared with the age of galaxies.

5.1.2.1 H–R Diagram

The Hertzsprung–Russell (H–R) diagram indicates the relationship between stellar luminosity as compared to our sun and the effective radiating temperature of the star (Fig. 5.1). It can be used to categorize different types of stars. Of all stars, 90% fall in the band running from the upper left to the lower right region in the diagram. This band is referred to as the main sequence. It consists of "normal stars". The sun occupies a spot near the middle of the main sequence. The sun is expected to evolve off the main

Fig. 5.1 Hertzspring–Russell (H–R) diagram showing different classes of stars (Rollinson 2007). The *arrows* show a typical birth to death cycle of a small star (*lower part of diagram*) and of a large star with a mass 25 times that of the sun (*upper part of diagram*)

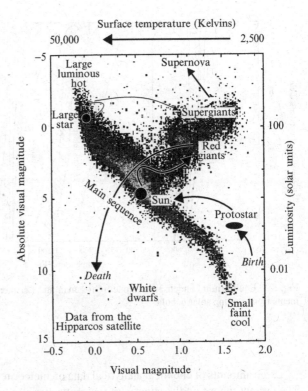

sequence and become a red giant (shown at the upper right in the diagram) and eventually a white dwarf (at the lower left). Red giants are cooler, but more luminous than stars on the main trend. The stars distributed on the left side of the main sequence region are hot, small, and very dark. These are called white dwarfs.

5.2 Formation of the Protosolar System

At the final evolutionary stage of stars, giant and supergiant stars explode. This explosion is accompanied by a brightening to more than one hundred million times its previous luminosity. The phenomenon is called a supernova. In the explosion, gravitational collapse occurs rapidly in the star's core, sometimes resulting in the formation of a neutron star. If the gravitational force of the star is very strong, light cannot escape from inside it. This is called a black hole.

The explosion of a supernova near the protosolar system galaxy is thought to have caused the mixing of the exploded substances with interplanetary dust particles concentrated in the protosolar system galaxy and triggered the formation of the solar system. This is based on evidence like the existence of large amounts of ^{26}Mg in the Allende meteorite discovered in Mexico.

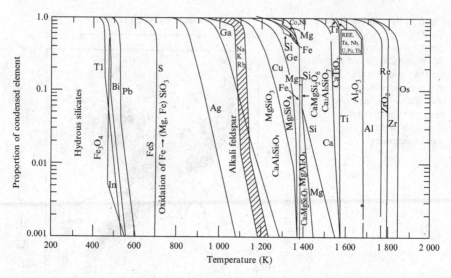

Fig. 5.2 Equilibrium condensation model (Grossman and Larimer 1974). Condensation of elements from cooling solar nebula

Large amounts of chemical analytical data on meteorites have been accumulated. The current theory on the genesis of the protosolar system was proposed based on these data. For example, Urey (1952) proposed a chemical equilibrium model, which was followed by proposals of several other chemical models. These models are subdivided into reduction and condensation models. Reduction models note that various meteorites were formed by oxidation–reduction reactions such as "FeO (iron oxide)$\rightarrow$Fe (iron)$+1/2O_2$ (oxygen), where "FeO" is the FeO component in minerals. If this reaction proceeds from the left hand side to the right, FeO is converted to Fe and O_2. If coalescing particles contain "FeO" aggregate, a stony meteorite forms. Oxidizing and reducing substances are necessary for the reaction above to proceed, but there are no such substances in the universe. It is thought that a significant change in temperature may also cause the reaction discussed above. An uneven temperature distribution in the protosolar system might have caused the diversity of meteorites we observe. The primordial galaxy could have collapsed suddenly. It entered the Hayashi phase which is a very brilliant stage, 10^3 times as bright as the sun. Tauri stars are in this stage, and so it is referred to as the T-Tauri stage. In such high temperature conditions, FeO changes to Fe and O_2 without the presence of reducing substances. After this stage, the protosun and protogalaxy shrank in size and brightness, and the temperature of the protosun increased. It took less than a hundred million years from the rapid collapse to go through the Hayashi phase and began the sun's hydrogen nuclear fusion reaction.

Equilibrium condensation models detail the sequence in which elements condensed into solid phases as the hot protogalaxy cooled (Fig. 5.2). Os, Zr, Re, Ti, Al,

REE (rare earth elements), and Fe condense at high temperatures. Pyroxene, olivine, and feldspar precipitate at intermediate temperatures. Sulfides, Pb, Bi, Tl, iron oxide (Fe_3O_4), hydrous silicates, and carbonates precipitate at low temperatures. The calculation of the equilibrium condensation process and the mineral compositions of meteorites indicate that iron meteorites and carbonaceous chondrites formed at high and low temperatures, respectively. The conditions under which various types of meteorites formed are estimated based on mineralogical and geochemical data from meteorites and calculations on condensation. Not only chemical reactions such as condensation and oxidation–reduction reactions, but physical processes like the dynamics of fine particles, collisions and accumulation are important in meteorite formation and in the production of a diversity of meteorites.

It is widely accepted that the sun, planets in the solar system and meteorites formed from a rotating, flattened disk of gas and dust. It is thought that various fine particles accumulated after condensing from their gaseous state. Various gases formed due to changes in temperature and oxidation–reduction conditions.

There are two main theories on the formation of the planets;

1. After the formation of fine particles, planets, and meteorites formed. This is the low temperature gas and dust theory.
2. Planetesimals about 10 km in diameter with masses of up to about 10^{15} kg accreted, forming planets. This is called the planetesimals theory.

It is now thought from computer simulations that the planetesimals theory is more plausible than the low temperature gas and dust theory.

Figure 5.3 is a schematic representation of the process from the cooling of the protosolar system to the formation of the solar system. The protosun grew due to the precipitation of gaseous components. The luminosity of the protosun increased by converting the kinetic energy of precipitating gaseous components, and the temperature and absolute magnitude moved to the upper left region on the H–R diagram. After that, the temperature of the protosun decreased as the precipitation of gaseous components decreased, leading to a change in the position toward the lower right region of the H–R diagram. The time required for this process was about a hundred thousands years. When the gaseous clouds disappeared, the sun entered the T-Tauri stage. A T-Tauri star is defined as one that does not belong to the main sequence of normal stars and red giants, and is brighter and cooler than the sun.

In the 1960s and 1970s, Dr. Chushiro Hayashi and his coworkers indicated that the sun, belonging to the main sequence, was formed by the collapse of a T-Tauri star (Hayashi et al. 1979). With this gravitational collapse and the subsequent increases in the T-Tauri star's density and core temperature, deuterium nuclei were formed by the collision of hydrogen nuclei. The density increased further and the temperature increased to more than one million degrees Celsius, and the hydrogen nuclear fusion reaction began. The shrinking ended when nuclear fusion and gravitational forces became balanced, after which the size does not change. Then, when the hydrogen in the core is consumed, leaving only helium, no nuclear fusion will occur, resulting in a collapse, accompanied by the generation of heat, raising the temperature. Nuclear fusion reactions will occur again and the sun

Fig. 5.3 Schematic representation of the process from the cooling of protosolar system to the formation of solar system (Matsui 1990)

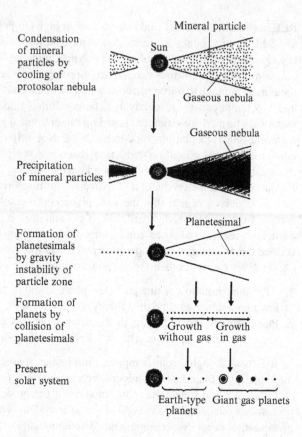

Condensation of mineral particles by cooling of protosolar nebula

Precipitation of mineral particles

Formation of planetesimals by gravity instability of particle zone

Formation of planets by collision of planetesimals

Present solar system

will expand, resulting in a red giant. When the sun is 12.3 billion years old (7.7 billion years older than now), it will become a white dwarf, signaling the end of the sun's life.

5.3 Planets in the Solar System

5.3.1 *Comparative Planetology*

Recent expedition to some of our solar system's planets has provided us with large amounts of information on the planets and the earth's moon. This information is useful in constraining our understanding of the origin and evolution of the earth. Comparative planetology is important for the following reasons: (1) Continuous changes in the earth's surface environment have been caused by weathering and erosion by the atmosphere and water. However, records of ancient times are preserved in the other earth-type planets because almost no liquid water flows there and there is little free oxygen in the atmosphere. (2) Plate tectonics is actively operating in the earth but not in the other planets and the moon. Plates subduct to the interior of the earth, and so no truly ancient records of the earth's surface have been preserved.

5.3.2   The Earth's Moon

The important characteristic features of the earth's moon are as follows:

1. The moon has a crust about 100 km thick. Feldspar is the major constituent of its crust, which is about 4.5 Ga old.
2. It is thought from the many craters on the surface of the earth's moon that bolide bombardment occurred around 0.5–1.0 hundred million years after the crust formed. This suggests that the heavy bombardment occurred in the earth's initial stage. Craters formed during that period on the earth might have eroded and weathered due to the effects of surface water containing oxygen and the atmosphere.
3. No atmosphere or liquid water are present on the lunar surface, and no tectonic or volcanic activities are observed on the moon.
4. The earth's moon is characterized by a layered structure somewhat like that of the earth. The upper mantle of the moon is composed of olivine and pyroxene, similar to earth's mantle. The presence of a core is not certain. If one exists, it is less than 400 km in radius.
5. The lunar "ocean" occupies 17% of its total surface area. This ocean is composed of basalt that is 3.2–3.9 Ga old.

The following hypotheses have been proposed to explain the genesis of the moon (Fig. 5.4): (1) the giant impact model, (2) the fission (parent earth and child moon) model, (3) the co-accretion model, in which the earth and moon are brothers, and (4) the capture model. Among these hypotheses, (1) is the most plausible (Stevenson 1987). The other hypotheses are either physically implausible or would produce a lunar composition different from analytical data available about lunar rocks. It is thought that the origin of the moon is an oblique impact between the proto-earth (probably molten) and a Mars-sized planetesimal about 15% the mass of the earth. The debris from the impact reassembled in orbit around the earth to form the moon. Subsequently, a magma ocean 100 km thick formed on the moon. Plagioclase crystallized in the magma ocean, ascended, and formed a global surface layer of anorthosite. Tungsten isotope data on lunar rocks indicate that the moon formed about 30 Ma after the formation of the solar system (Schoenberg et al. 2002).

5.3.3   The Earth-Type Planets

The earth-type planets include Mercury, Venus, and Mars. Their bulk density is high compared with the other planets in solar system, indicating they are mainly composed of solid materials.

Fig. 5.4 Hypotheses for origin of the moon (Broecker 1988)

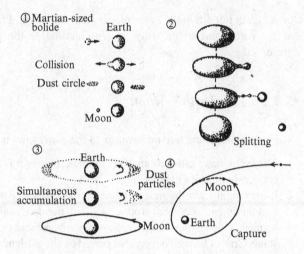

5.3.3.1 Mercury

Mercury's atmosphere contains very small amounts of Ar, Ne, and He. These components are supplied by the solar wind. Many impact craters are observed on its surface. The age of these craters ranges within 3–4 Ga, indicating no tectonic activity in Mercury since its formation. The atmospheric temperature varies widely in a range from −200°C to +500°C. Mercury's bulk density is high at 5.44 g/cm^3, indicating the presence of a core composed of Fe and Ni. The core is estimated to be 1,600 km in radius.

Like earth's moon, no tectonic activity has occurred on Mercury due to its small size. It is thought that the high density is due Mercury's relatively large core. The core is thought to be so large for the planet's size because of the loss of mantle composed of silicates. A bolide in the size range of the Moon to Mars may have caused the loss of Mercury's mantle.

5.3.3.2 Venus

The surface of Venus is covered by a thick, CO_2 rich (97%) atmosphere containing argon. A sulfuric acid cloud layer is present 50 km above the surface. Due to CO_2's greenhouse effect, the surface temperature is high, about 730 K. Venus's surface shows topographic features caused by volcanic activity. The topography has features similar to folded mountains, ridges, and rifts, suggesting that plate tectonic motions took place during the past. However, there are also ring depression structures several hundreds of kilometers in diameter, suggesting that plume tectonics were dominant rather than plate tectonics. The ages of rocks on the surface are estimated to be about 4.5 Ga from the density and distribution of craters. The size and composition of Venus is similar to the earth. Thus, comparisons between Venus and the earth constrain the genesis of the earth. No liquid water is present on Venus and the high partial pressure of CO_2 in Venus's atmosphere (P_{CO2}) is different from the earth. The high partial pressure of CO_2 is thought to be because atmospheric CO_2 has not been removed since Venus has no oceans where CO_2 could have reacted to form carbonates. We infer that the P_{CO2} of the protoearth's atmosphere was similar to that of Venus. The layer structure of Venus's interior is thought to be similar to the earth but this is not clear.

5.3.3.3 Mars

The Martian atmosphere consists mainly of CO_2 (95%) and small amounts of N_2, Ar, O_2, and water. Many craters are observed on the surface, particularly in the southern hemisphere, and volcanic topography is common, particularly in the northern hemisphere. The composition of the surface soil is SiO_2 44.7%, Al_2O_3 5.7%, Fe_2O_3 18.2%, MgO 8.3%, CaO 5.6%, K_2O <0.3%, TiO_2 0.9%, SO_3 7.7%, and Cl 0.7%.

This composition is similar to basalt. However, in addition to basaltic material, salts such as sulfates and chlorides may exist. Mars has a mantle and core. These are richer in iron and denser than their counterparts on earth.

A huge shield volcano named Mount Olympus, which is 600 km in diameter and 25 km high, and canyons characterize the surface topography. No such volcano exists on the earth's surface. This is probably due to the plate motion on the earth causing continuous changes of volcanic eruption sites. Valley Marines, the largest canyon on the surface of Mars, is 4,000 km long. It is thought to have been formed by the effects of liquid water flow. The origin of liquid water on Mars is controversial, but it seems likely that ice near the surface melted due to meteorite impact and/ or volcanic eruption.

The Viking Mars expedition in 1976 found topographies caused by river flooding and flow in the highlands of the Southern Hemisphere. This indicates that a warm climate existed in the past. "Opportunity", NASA's Mars Pathfinder rover, found evidence of the presence of large amounts of liquid water on the surface of Mars. It found iron oxide (hematite) and sulfate (jarosite), which form only in the presence of liquid water. Rocks with sedimentary structures were also discovered. Therefore, we are certain that large amounts of liquid water existed on the surface of Mars. However, the origin of that water is uncertain. Seawater and lakewater are possible sources. Jarosite is formed by the reaction of sulfuric acid solutions of volcanic origin with rocks in lake environments on the earth. Thus, a lakewater environment seems likely at the site the jarosite precipitated on the Martian surface. Generally, such an environment, warm with liquid water present, is necessary for the origin of life, and so evidence of the presence of liquid water on the Martian surface in the past suggests that life may have existed on Mars.

5.3.4   The Giant Planets

The Giant planets consist mainly of gases and solids. The gas compositions of
Jupiter and Saturn are similar to that of the sun. Their atmospheres are mainly composed of hydrogen and helium. The average cloud temperatures of Jupiter and
Saturn are −150°C and −180°C, respectively. Saturn's interior consists of rock containing SiO_2, MgO, Fe, Ni, etc. with a mass 10–15 times that of the earth, an ice
core, and a metallic hydrogen mantle. Jupiter and Saturn have heat sources in their
interiors and radiate considerable amounts of energy. There is an active volcano on
one of Jupiter's satellites, Io. The surface of Europa, another of Jupiter's satellites,
is covered by thick ice. Liquid water is thought to be present underground on
Europa. Igneous activity likely melted the ice underground. There is conjecture that
life may be present in Europa's ocean (Naganuma 2004). Volcanic activity and the
presence of liquid water and life may also be possible in the "ice giants", Ganymede
and Callisto, two more of Jupiter's satellites.

Uranus and Neptune are giant planets like Jupiter and Saturn. They are 80% icy
water and methane by mass. The main component in the atmosphere of Titan,
Saturn's largest moon, is N_2 with minor amounts of Ar, H_2, CO, CO_2, H_2O, and
hydrocarbons (C_2H_6, C_2H_2, etc.). Triton is the largest satellite in the solar system and
is larger than Pluto. Triton's surface temperature is about −220°C, which is the cold-
est in the solar system. Triton's surface is dominated by frozen ices of several types,
probably methane, ammonia, carbon dioxide, and nitrogen.

Pluto has a solid surface, which is different from the giant planets. Its atmospheric pressure is very low. The atmosphere is probably composed of methane
evaporated from the surface ice, CO, and nitrogen. Recently, icy methane was discovered. Its surface temperature is −230°C, and its bulk density is 2.03 g/cm^3.
Organic matter, including methane, exists on the surface. The interior is composed
of ice and rocks. Pluto is a large bolide in the Kuiper Belt. Recently, it was excluded
from the list of planets in the solar system, mainly because of its small size compared with the other solar planets.

5.3.5   Asteroids, Kuiper Belt Objects (KBOs), and Comets

Asteroids are distributed between the orbits of Mars and Jupiter. There are more
than 14,000 asteroids whose orbits have been observed. Asteroids are very dark,
with an average magnitude dimmer than 6.8. Most asteroids are distributed between
the orbits of Mars and Jupiter, but some cross inside earth's orbit. Impact craters
caused by asteroids have been discovered on the earth's surface, such as the crater
left by the asteroid impact in Siberia in 1908. It is likely that an asteroid impact
caused the Mass Extinction, when dinosaurs became extinct at the Cretaceous/
Tertiary boundary (see Sect. 6.2.8). The asteroids that arrive on the earth and Mars
are rocks containing iron. They are aggregates of small fragments of planetesimals
from the primordial solar system, rather porous and not very dense. Asteroids that

impact Jupiter are composed of dark or reddish substances, which are remnants from before the birth of the planets.

Since the first discovery of the Kuiper Belt objects (KBOs) in 1992, the number of identified KBOs has increased to over a thousand, and more than 700 KBOs over 100 km in diameter are believed to exist. Pluto and its moon Charon are large bodies in the Kuiper Belt.

Comets and asteroids are surviving planetesimals from the solar nebula, preserved building blocks of the planets. Comets are considered to be relic materials from the outer regions of the solar nebula. They are composed of materials similar to interplanetary gas and dust particles (C_2, CN, CH, OH, NH_2, CO^+, CO_2^+, N_2^+, etc.).

5.3.6   Characteristic Differences Between the Earth and Other Planets

The main characteristics that differ between the earth and the other planets are summarized as follows.

1. The earth's atmosphere is characterized by high concentrations of N_2 and O_2, and a low concentration of CO_2. The atmospheres of the other planets are characterized by the presence of CO_2, He, and H_2.
2. There is liquid water on earth, but not on the other planets, although it was present on Mars in the past. Solid water (ice) is abundant in the other planets, particularly the giant planets. The earth is the only planet in the universe that is known to habitable for life.
3. The frequency distribution of earth's topographic height is bimodal, but on Venus, Mars, and earth's moon, it is unimodal (Fig. 5.5). The earth's bimodal distribution is due to the crust being divided into continental (mostly granitic) and oceanic (mostly basaltic) types.
4. Giant volcanoes like Mount Olympus on Mars exist on the other terrestrial planets.
5. Plate tectonics operates on the earth. Features 3 and 4 relate to plate tectonics on the earth. No evidence of plate tectonics is found on the other planets.
6. Plume tectonics was dominant on Mars and Venus, indicating volcanism on these planets was caused by plume tectonics.
7. At present no liquid water exists on any planet other than the earth. The conditions for the presence of liquid water on the planets largely depend on temperature, which is mainly governed by the distance from the sun and planet size, which controls the gravity. On earth, these conditions were the most favorable for the presence of liquid water of all the planets. It is thought that the presence of liquid water on the earth throughout its history has been determined by its thermal history and its evolution after its birth as a planet.

Oceans are present on earth. Protoearth's atmospheric CO_2 pressure was probably very high. After the formation of the oceans, the atmospheric CO_2 dissolved into

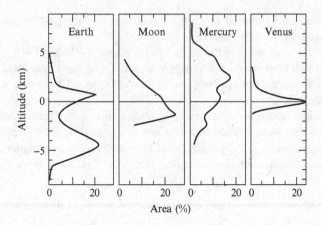

Fig. 5.5 Distribution of elevation of earth-type planet (Masursky et al. 1980)

the ocean water. Weathering of silicate rocks by interaction with CO_2-containing aqueous solutions (the oceans and terrestrial water) leads to high concentrations of Ca^{2+}, Mg^{2+} and HCO_3^- in the oceans. Carbonates precipitated from these aqueous solutions. Due to this precipitation, atmospheric CO_2 was reduced significantly, decreasing the P_{CO2}. In contrast to the earth, Venus's atmospheric CO_2, which is very concentrated at about 100 atm, did not decrease because Venus had no ocean. Life could have originated in the earth's ocean, but is unlikely to have originated on the other planets because of the lack of oceans, except perhaps on Mars.

If magma is generated under hydrous conditions, the magma becomes more felsic (granitic) compared with the source rocks. Hydrous conditions deep in the interior of the earth are caused by water carried by subducting slabs containing alteration minerals. Therefore, it could be said water and plate tectonics caused the character of the earth.

There is considerable mass transfer in the earth's surface environment because of the water cycle. Ore deposits formed by these processes are used as mineral resources. Fossil fuels such as coal and oil are not formed without water and organisms. Most nonrenewable resources form in relation to the water cycle. Mass cycling is occurring in the interior of the earth through plate motion, plume motion and mantle convection. Therefore, it can be said that entire earth is very dynamical.

5.4   Chapter Summary

1. Our universe began from the big bang and elements, stars and galaxies formed irreversibly and then evolved.
2. In the rotating disk of the protosolar nebula, fine grains condensed, aggregated, and grew, leading to the formation of planetesimals. The planetesimals collided and accumulated, forming planets, including the earth in our solar system.

3. The planets in the solar system include the earth-type planets Mercury, Venus, Earth, and Mars and the Giant planets Jupiter, Saturn, Uranus, and Neptune.
4. There are planets, moons, asteroids, comets, and Kuiper Belt objects (KBOs) like Pluto in the solar system.

References

Broecker WS (1988) How to build a habitable Earth. Kodansya, Tokyo (in Japanese)

Grossman L, Larimer JW (1974) Early chemical history of the solar system. Rev Geophys Space Phys 12:71–101

Hayashi C, Nakazawa K, Mizuno H (1979) Earth's melting due to the blanketing effect of the primordial dense atmosphere. Earth Planet Sci Lett 43:22–28

Hubble EA (1929) A relation between distance and radial velocity among extragalactic nebulae. Proc Natl Acad Sci USA 15:168–173

Masursky H, Eliason E, Forol PG, Mcgill GE, Pettengill GH, Schaber GG, Schabert G (1980) Pioneer Venus radar results: geology from images and altimetry. J Geophys Res 85:8232–8260

Matsui T (1990) The newest Earth science. Kodansya, Tokyo (in Japanese)

Naganuma T (2004) Europe, life star. NHK Books, Tokyo (in Japanese)

Penzias AA, Wilson RW (1965) A measurement of excess antenna temperature at 4080Mc/s. Astrophys J 142:419–421

Rollinson H (2007) Early earth system. Brackwell

Schoenberg R, Kamber BS, Collerson KD, Eugster O (2002) New W-isotope evidence for rapid terrestrial accretion and very early, core formation. Geochim Cosmochim Acta 66:3151–3160

Stevenson DJ (1987) Origin of the moon—the collision hypothesis. Annu Rev Earth Planet Sci 15:271–315

Urey HC (1952) The planets. Yale University Press, New Haven

Chapter 6
Evolution of the Earth System

The origin and evolution of the earth system have been deciphered by various methodologies including (1) geochronology, (2) comparative planetology, (3) computer simulation, (4) isotope geochemistry, and (5) geology of past geologic events. Based on these methods, the following important topics concerning the origin and evolution of the earth system have been debated and elucidated: (1) What is the origin of earth? (2) Was primordial earth cold (the cold origin theory), or hot (the fireball origin theory)? (3) Was earth formed by heterogeneous accretion or homogeneous accretion? (4) When was the layered structure of the earth formed? (5) Were the atmosphere and oceans formed by rapid initial degassing or continuous degassing from the interior of the earth? (6) Were the primordial atmosphere and oceans reducing or oxidizing? (7) How did the atmosphere and oceans evolve?

These subjects will be discussed below.

Keywords Climate change • Evolution of earth system • Geochronology • Mass extinction • Origin of life • Origin of the earth system

6.1 Geochronology

Earth's system has evolved from its initial stages to the present time. To understand the evolution of the earth system it is essential to know when various geologic events occurred and geologic materials formed throughout earth's history. Age determinations of constituents of the earth are possible based on geochronology. There are two types of methods for determining these ages: (1) relative age dating and (2) radiometric (absolute) age dating.

N. Shikazono, *Introduction to Earth and Planetary System Science*,
DOI 10.1007/978-4-431-54058-8_6, © Springer 2012

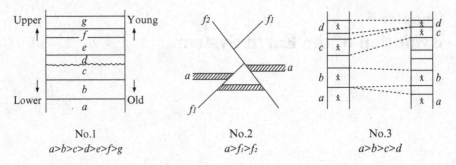

Fig. 6.1 Three laws estimating relative geologic ages (Hamada 1986). x > y means x is older than y

6.1.1 Relative Age Dating

Relative age dating can be made from (1) the stratigraphic record, (2) fossils as timepieces, and (3) the relationship between geologic bodies such as unconformities and cross-cutting relationships (Fig. 6.1).

We can interpret the relative ages of geologic events from sedimentary records based on the principle of superposition. It states that each layer of sedimentary rock is younger than the one beneath it and older than the one above it if the layers have not suffered tectonic disturbance.

Fossils, which are the remains of extinct life forms, can be used to date the relative ages of sedimentary rocks if the occurrence of certain fossils is limited to particular layers and formations of sedimentary rocks.

An unconformity is the boundary along which two existing formations contact each other. It is a surface between two layers that were not laid down in an unbroken fashion. Unconformities are created by uplift and erosion, followed by subsidence and another records of sedimentation (Siever et al. 2003).

If magma intrudes into sedimentary rocks and solidifies, the solidified magma (intrusive rocks) cut the sedimentary rocks. This cross-cutting relationship indicates that the magma intrusion occurred later than the sedimentation. Fault planes in geologic bodies also exhibit cross-cutting relationships. They indicate that faulting occurred later than the formation of the geologic bodies.

Using these three methods, paleontologists and sedimentologists investigated outcrops all over the world in the nineteenth and twentieth centuries and devised the entire geologic time scale (Fig. 6.2).

6.1.2 Radiometric (Absolute) Age Dating

Radioactive atoms decay with time as expressed by the following equation:

$$dN / dt = \lambda N$$

Fig. 6.2 The geologic time scale (Stanley 1999). The *numbers on the right* represent the ages of the boundaries between periods and epochs in millions of years. The Holocene Epoch (the past 10,000 years or so) is also known as the Recent

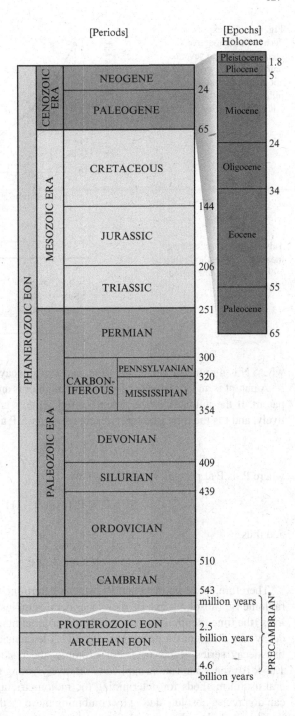

Fig. 6.3 The exponential decay of a radioactive element, showing several half-lives; each would be about approximately 1.5 billion years in the case of ^{40}K (Press and Sieber 1994)

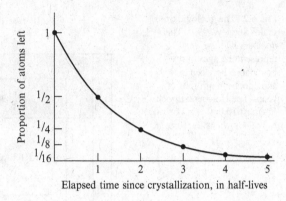

Table 6.1 Radiometric age measurement method (Ojima 1990). $\tau_{1/2}$: half-life period (year)

	$t_{1/2}$ (Year)
^{40}K $\rightarrow$ ^{40}Ar	1.25×10^9
^{87}Rb $\rightarrow$ ^{87}Sr	4.88×10^{10}
^{147}Sm $\rightarrow$ ^{143}Nd	1.06×10^{11}
^{176}Lu $\rightarrow$ ^{176}Hf	3.6×10^{10}
^{187}Re $\rightarrow$ ^{187}Os	4.56×10^{10}

where N is number of nuclei, t is time, and λ is a decay constant.

A parent is an atom that decays and its daughter is an atom made from part of the parent. If the numbers of the parents and daughters are denoted P and D, respectively, and t is the time from the present to the past, P at time t is given by

$$P = P_o \exp(\lambda t)$$

where P_o is P at present. D at t is given by

$$P - P_o = P_o \left(\exp(\lambda t - 1)\right)$$

and thus,

$$t = 1/\lambda \ln(1 + D/P_o)$$

Therefore, if we know D and P_o, we can determine t (Fig. 6.3) and so know the radiometric (absolute) age of rock or mineral samples. Decay constants and half-lives (the time required for one-half of the original number of radioactive atoms to decay) are different for different radioactive elements as described by the radioactive decay series (Table 6.1). For example, the residence time of ^{40}K $\rightarrow$ ^{40}Ar is 1.25×10^9 years and can be applied to relatively young geologic ages. We have several useful methods for determining the radiometric ages of geologic events, and can do precise absolute dating by combining these methods.

We can apply radiometric dating methods not only to ancient geologic events but also to recent geologic ones. For example, we can estimate the sedimentation rate of recent shallow sediments and the growth rate of manganese nodules on the seafloor using ^{222}Ra, ^{23}Th, ^{231}Po, ^{210}Pb, ^{234}U, and ^{10}Be. Radiocarbon (carbon 14) dating has been used to date materials back to 50,000 years ago. It is used in archeology and for reconstructing the paleoclimate and deciphering ocean circulation patterns.

6.2 Origin and Evolution of the Earth System

6.2.1 Formation of Early Earth (About 4.6–3.0 Billion Years Ago)

Generally, the geologic time scale is divided into four major time units: eons, eras, periods, and epochs. For example, there are three eons, the largest division of history: the Archean eon, 4 billion years ago to 2.5 billion years ago; the Proterozoic eon, 2.5 billion years to 543 million years ago; and the Phanerozoic eon, the past 543 million years. Division of the geologic time scale is usually based on paleontology and sedimentology. However, here we have divided it into three parts, mainly based on geochronological data to aid in understanding the earth's history: 4.6–3.0 billion years ago, 3.0–2.0 billion years ago, and 2.0 billion years ago–the present.

6.2.1.1 Homogeneous and Heterogeneous Accretions

It is generally accepted that the earth was formed 4.6–4.5 billion years ago from the evidence that the ages of most chondrite meteorites in the solar system, based on the isochron diagram showing lead isotope ratios, are 4.6–4.5 billion years old.

Two models to explain the mechanism of the formation of the earth have been proposed, the homogeneous accretion model and the heterogeneous accretion model. The homogeneous accretion model says that homogeneous bodies accreted to form the protoearth, while the heterogeneous accretion model says that heterogeneous bodies accreted. The evidence supporting the heterogeneous accretion model are as follows:

1. This model can explain the bulk and isotopic compositions of the earth.
2. The energy required for separating metallic iron and silicates is small if the metallic iron and silicates formed the protoearth at different stages.
3. It is difficult for metallic iron to accumulate in the core by homogeneous accretion. Iron has to be molten and the mantle consisting of silicates must also be molten or soft for the iron to be able to sink to the core of the earth.

The evidence supporting the homogeneous accretion model is as follows.

1. Most minor elements' partitionings between the core and mantle are in equilibrium. This indicates the separation of core and mantle at happened late in the formation of the earth.

2. Accretion of planetesimals with the compositions of various meteorites can explain the composition of the earth.

Despite a great deal of investigation, it is still uncertain which model is better for explaining the formation of the protoearth.

6.2.1.2 Earth's Source Materials

Several models explaining the bulk composition of the earth have been proposed. Ringwood (1979) proposed the C1 chondrite model for the origin of earth. C1 chondrites contain the highest amounts of volatile elements among the carbonaceous chondrites (C 3.5%, H_2O 20%, S 6%) and have elemental compositions similar to the relative abundance of elements in the atmosphere of the sun (Fig. 2.7). This model shows the reduction of iron oxide to metallic iron like

$$FeO + H_2 \rightarrow Fe + H_2O$$

$$FeO \rightarrow Fe + 1/2O_2$$

For this reaction to proceed, large amounts of O_2 are generated and H_2 or other reducing materials are required. C1 chondrites are considered to be the most undifferentiated primordial substance, and their REE (rare earth element) pattern is quite different from that of the earth. The earth's atmospheric xenon isotopic composition and krypton isotopic composition are also different from those of carbonaceous chondrites. The earth's chondrite-normalized REE pattern is similar to that of volatile-poor iron meteorites, indicating that the earth was built from volatile-poor meteorites. Earth's minor element contents cannot be explained only by the contents of carbonaceous chondrites. For example, mixing carbonaceous chondrites (formed at low temperatures with oxidizing material) with enstatite chondrites (formed at high temperatures with reducing material) can reasonably explain the Zn, In, Ce, and C contents of mantle of the earth. Therefore, the most plausible source material for the earth is planetesimals consisting of various meteorites including iron meteorites, carbonaceous chondrites, and enstatite chondrites. ·

6.2.1.3 Formation of the Atmosphere

We now consider how the earth's atmosphere was formed. The theory that the earth's atmosphere today originated from its primary atmosphere was denied by Brown (1949), who found that the REE abundances in earth's atmosphere today are different from solar nebula REE abundances. Therefore, it is clear that degassing from the earth's interior after the earth's formation is the major source of the early earth's atmosphere. In his 1951 paper titled "Geologic history of sea water—an attempt to state the problem—", Rubey proposed a "continuous degassing model" in which the volatile components have degassed gradually and continuously from earth's interior throughout geologic history. This model was widely accepted because it was consistent with the cold origin theory widely accepted at that time. It is based on the following geologic evidence: (1) There are fewer carbonate rocks of the older ages. (2) Magnesite ($MgCO_3$) is not found in any of the geologic ages. (3) The composition of excess volatile matter is similar to the gas composition in present-day volcanoes. The excess volatile matter is equal to the amount of volatiles in the atmosphere, oceans, and sediments minus the amount of volatiles produced by the weathering of rocks.

The rapid degassing model, which posits that degassing occurred during the several billion years after the formation of the earth, is more plausible based on the following evidence:

1. The $^{40}Ar/^{36}Ar$ ratios of the present-day atmosphere and upper mantle are 295.5 and 20,000, respectively. Such a high value in the upper mantle can be explained by rapid initial degassing of ^{36}Ar, which is a stable Ar isotope. Thus, its concentration does not change over time. In contrast, the concentration of ^{40}Ar in the mantle increases over time because it is a product of the decay of ^{40}K, which is contained in the mantle but not in the atmosphere.
2. Xenon isotope data also indicates initial rapid degassing.
3. Computer calculations indicate that if the protoearth was formed by the accretion of planetesimals, a "magma ocean" could have been formed, accompanied by the degassing of CO_2 and other gases and the formation of a protoatmosphere.
4. Oxygen and hydrogen isotope studies on the surface water existing on earth's surface now clearly show that the water cycle is prevailing on and near the earth's surface. This data, and the fact that no juvenile water is found in the earth today, suggest no continuous degassing of juvenile water from the earth's interior throughout geologic history. This supports the idea of rapid degassing during the initial stage of the earth's formation. However, helium isotope data indicate that helium is degassing from the earth's interior (mantle) even at the present time. Therefore, not all volatiles degassed during the initial stage, and some, like helium, have been degassing since the initial stage to the present.
5. Evaporites form by the evaporation of seawater, leading to increased salinity and the precipitation of salts. The sequence of minerals in evaporite reflects the seawater chemistry at the time of evaporite formation. Holland (1972) published the paper, "Geologic history of seawater—an attempt to solve the problem," and he found systematic sequences of minerals in the evaporite from various geologic ages. These sequences are systematic and consistent in evaporites with different ages. He reasoned that seawater chemistry with regard to the constituents of evaporates, which include alkali elements, alkali earth elements, etc. has been constant, which disagrees with the continuous degassing model by Rubey (1951).

6.2.1.4   Variation of the Compositions of Seawater and the Atmosphere

What was the composition of the proto-atmosphere when it was formed by degassing from the interior? It has been proposed that the partial pressure of CO_2 (P_{CO_2}) was very high and no free oxygen gas was present in the earth's initial stages. The reasons P_{CO_2} was thought to be high are as follows:

1. Sunlight is generated by nuclear fusion of hydrogen to helium. As this reaction proceeded, temperature and pressure increased gradually, suggesting that the surface temperature of the early earth was lower than it is today. However, various lines of evidence suggest the surface temperature at the earth's formation was higher than it is today. This apparent discrepancy is called the "faint young sun paradox". The most likely solution is that a significantly high concentration of greenhouse gas caused the high temperature on the earth's surface. For example, if P_{CO_2} at that time was very high (around 100 atm), the temperature should have been high enough to melt the ice sheets.
2. Studies on the oxygen isotopic compositions of chert and carbonates of Archean age indicate that the temperature of the Archean ocean was high, probably caused by the greenhouse effect of atmospheric CO_2.

However, recent investigations of the chemical properties of the proto-atmosphere suggest that the main greenhouse gas was methane rather than CO_2 (Shikazono 1997). Shikazono (1997) indicated the rapid removal of atmospheric CO_2 due to the atmosphere–ocean–oceanic crust interaction soon after the formation of the oceans was caused by the following reactions:

$$Ca^{2+} + 2HCO_3^- \rightarrow CaCO_3 + CO_2$$

$$CO_2 + H_2O \rightarrow HCO_3^- + H^+$$

Ca^{2+} in the above reaction is derived from the oceanic crust via a seawater–oceanic crust interaction. Because of this reaction, the atmospheric CO_2 concentration decreased to a level similar to the present-day level over a short period. Such a low CO_2 cannot explain the "faint young sun paradox". A different possible greenhouse gas to consider instead of CO_2 is methane. Methane could be generated by methanogenic bacteria or methanogenic reactions like

$$CO_2 + 4H_2 \rightarrow CH_4 + 2H_2O$$

and also by inorganic reactions like

$$FeO + CO_2 + 3/4H_2 \rightarrow 1/2Fe_2O_3 + CH_4 + 3/2H_2O$$

Recent geophysical and geochemical studies on earth-type planets in the solar system and on the earth's moon indicate the presence of a magma ocean on the protoearth.

Table 6.2 Comparison of calculated fugacity ratios in gases at $f_{O_2} = 10^{-12}$ and 10^{-8} with observed fugacity ratios in same present-day volcanic gases (Holland 1984)

Fugacity ratio	$f_{O_2} = 10^{-12.0}; f_{H_2O} = 5$atm		$f_{O_2} = 10^{-8.0}; f_{H_2O} = 5$atm		Volcanic gases from Surtsey, Hawaii, and Erta'Ale
	$T = 1,400$ K	$T = 1,500$ K	$T = 1,400$ K	$T = 1,500$ K	
f_{H_2} / f_{H_2O}	$10^{-0.3}$	$10^{+0.3}$	$10^{-2.3}$	$10^{-1.7}$	$10^{-2.1} - 10^{-1.3}$
f_{CO} / f_{CO_2}	1.0	$10^{+0.7}$	$10^{-2.0}$	$10^{-1.3}$	$10^{-1.5} - 10^{-1.3}$
f_{CH_4} / f_{CO_2}	$10^{-3.5}$	$10^{-2.5}$	$10^{-11.5}$	$10^{-10.5}$	No CH$_4$ observed
f_{H_2S} / f_{SO_2}	$10^{+3.4}$	$10^{+4.7}$	$10^{-2.6}$	$10^{-1.3}$	No H$_2$S observed
$f_{NH_3} / f_{N_2}^{1/2}$	$10^{-3.6}$	$10^{-2.8}$	$10^{-6.6}$	$10^{-5.8}$	

Chemical equilibrium between the magma and gas was attained in a short period after the formation of the earth. If metallic iron (Fe) and iron oxide (FeO) were in equilibrium with respect to the reaction

$$3Fe + 3/2O_2 = 3FeO$$

we can estimate the oxygen fugacity (f_{O_2}), given temperature and total pressure. Considering equilibrium for the reaction between olivine, quartz, magnetite, and gaseous species, f_{O_2} can be estimated from the equilibrium equation for the reaction.

$$3/2Fe_2SiO_4 + 1/4O_2 = Fe_3O_4 + 3/2SiO_2$$

We can estimate $f_{SO_2} / f_{H_2O}, f_{CO_2} / f_{CO}$, and f_{CH_4} / f_{CO_2} from the equilibria between minerals (iron oxides, silicates, and sulfides) and gaseous species. The equilibrium calculations indicate that $_{O2}f$ and which gaseous species dominate depend largely on the mineral species present. For example, the dominant species of sulfur gas (H$_2$S or SO$_2$) is different for different minerals such as metallic iron, magnetite, and olivine. We assume that the composition of gas in equilibrium with olivine and magnetite is the same as the proto-atmospheric gas composition. In this case, in general, P_{SO_2} exceeds P_{H_2S} (Table 6.2). If the temperature of such a gas decreases, SO$_2$ may change to H$_2$S in the reaction below.

$$SO_2 + H_2 \rightarrow H_2S + O_2$$

How much of the SO$_2$ changes to H$_2$S depends on the degree of the reaction between iron minerals and gaseous species (SO$_2$, O$_2$, and H$_2$), the supply rate of H$_2$ from the earth's interior and the release rate of H$_2$ from the atmosphere. In general, f_{SO_2} is higher than f_{H_2S} in volcanic gas in equilibrium with basaltic magma today (Table 6.2). However, this depends on the kind of basaltic magma. The f_{O_2} of gas in equilibrium with metallic iron is low, and f_{SO_2} is less than f_{H_2S}. It is likely that the volcanic gas was in equilibrium with metallic iron, which settled to the bottom of the magma ocean during that time.

It is thought that the earth's temperature decreased over time due to a decrease in heat generated by the collision of planetesimals with the protoearth, a decrease in the heat supply from the interior, and decrease in the concentration of the greenhouse gases CO_2 and CH_4 in the atmosphere.

When the surface temperature had decreased considerably, gases condensed and the oceans was formed. It seems likely that reduced sulfur species, H_2S, and CO_2 and CH_4 were dominant among the dissolved sulfur and carbon species.

The atmospheric composition at that time can be estimated based on the equilibrium calculations and the excess volatiles composition published by Rubey (1951). The pH of the condensed excess volatiles is 0.3, implying strongly acidic conditions. Such strong acid solutions interacted with basalt, and rapid neutralization occurred by the reaction.

$$\text{silicate} + H^+ \rightarrow \text{cation} + \text{clay mineral}$$

This reaction is very rapid. Thus, it is likely that the pH of the primordial ocean was nearly neutral, and the alkaline and alkaline earth element concentrations were not much different from those of the present day (Shikazono 1997).

When was the primordial ocean formed? The age of the metamorphic rocks in Isua, Greenland is 3.8 Ga. This metamorphic rock is one of the oldest rocks on the earth. The rocks from which the metamorphic rocks in Isua were formed were marine sedimentary rocks. Therefore, we are able to estimate the geochemical properties of the oceans and atmosphere after 3.8 Ga, based on the features of sedimentary rocks. In contrast, we have not found rocks from before this age, and so we cannot estimate the properties of the atmosphere and oceans earlier than that. Recently, zircon dated at 4.2–3.9 Ga was discovered in metamorphic rocks in Australia and Canada. However, it is impossible to estimate the environmental conditions at that time from only the zircon data.

The water–rock interaction thought to control ocean chemistry is related to magmatic intrusion at midoceanic ridges and ocean floor spreading. Riverine input to oceans is controlled by the continental area and distribution of continents, which are also related to tectonics. It is generally thought that there were no large continents when the primordial ocean was in existence. Therefore, it is likely that the oceanic basalt–seawater interaction was more important for controlling ocean chemistry than was riverine input.

The ocean at that time was warm, indicated by oxygen isotopic studies on chert. It is also thought that the temperature of the earth's interior was higher than it is today. One item of evidence supporting the warm earth is the presence of komachiite (basaltic rocks distinguished by the presence of higher magnesian ultramafic lavas) of that age, which suggests a higher geothermal gradient than today.

The tectonics at that time were considerably different from today. The geothermal gradient was higher, caused by the higher heat supply from the earth's interior. The lithosphere was thinner and the oceanic crust was thicker (about 25 km) than today. In such conditions, there were more frequent collisions of continents but subduction did not occur.

External systems also exerted considerable influence on the earth system. There was an impact crater stage at 4–3 Ga experienced by the earth's moon. Thus, it is thought that the earth also experienced an impact crater stage as the same time.

Fig. 6.4 Variations in geochemical features of sedimentary rocks with time (Veizer 1976). (**a**) K_2O/Na_2O of basement and sedimentary rocks. (**b**) $^{87}Sr/^{86}Sr$. (**c**) Europium anomaly

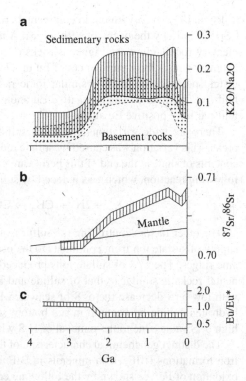

If planetesimals and meteorites impacted the earth frequently, large continents might have broken up into microcontinents. The continents were also separated by mantle plume intrusions.

The next question is what was the earth's interior like? It is known that paleomagnetism was very high at 3.5 Ga. Also, metamorphic rocks transformed from marine sediments from 3.8 Ga have been discovered in Greenland. These lines of evidence indicate that mantle was separate from the core and that the layered structure of the earth (core, mantle, crust, oceans, and atmosphere) had been established earlier than that.

6.2.2   From 3.0 to 2.0 Billion Years Ago

Various geochemical features of earth's surface environment that existed during this period have changed. For example, chemical compositions such as the K_2O/Na_2O ratio and the Eu anomaly, and isotopic compositions such as the $^{87}Sr/^{88}Sr$ ratio and the $\delta^{34}S$ of sedimentary and igneous rocks has changed greatly (Fig. 6.4). The K_2O/Na_2O ratio and the $^{87}Sr/^{86}Sr$ ratio both increased. The data suggest rapid growth of continents and the separation of the continental crust from the mantle. Continental crust is granitic in composition, containing large amounts of alkaline elements such as Rb and K. Rb enrichment in continental crust caused the increase in ^{87}Sr and the ^{87}Sr/^{86}Sr ratio by the decay of ^{87}Rb to ^{87}Sr. A negative Eu anomaly is found in sedimentary rocks (chert) and igneous rocks of this period. We think the Eu negative anomaly comes from enrichment of Eu in feldspar. Eu^{2+} can be substituted for Ca^{2+} in feldspar because of their similar ionic radii and charges. Magma from which feldspar fractionated has a negative Eu anomaly, while igneous rocks enriched in feldspar have positive Eu anomalies.

There are wide variations of δ^{34}S of sulfides like pyrite (FeS$_2$) in sedimentary rocks. This variation was caused by sulfate reducing bacteria. The sulfur and carbon contents of shale at the end of this period are positively correlated. This is due to the following reaction, which was induced by sulfate-reducing bacteria.

$$SO_4^{2-} + 2H^+ + CH_4 \rightarrow CO_2 + H_2S + 2H_2O$$

The other factors controlling the sulfur isotopic composition of seawater are the supply of sulfate ion from continents to the oceans and seawater cycling at midoceanic ridges. The δ^{34}S of sulfate ions produced by the oxidation of sulfide in continental rocks is similar to that of sulfide and is relatively low. Inputs of riverwater with low δ^{34}S decrease the δ^{34}S of seawater. However, reduction of sulfate ions to hydrogen sulfide by bacteria in sea bottom sediments causes sulfate ions to have high δ^{34}S values due to the removal of H$_2$S with low δ^{34}S.

The distinct geochemical characteristic of this period is the occurrence of banded iron formations (BIF). Iron minerals in BIF are considered to have formed by the oxidation of Fe^{2+} as shown by the following equation.

$$Fe^{2+} + H^+ + 1/4O_2 \rightarrow Fe^{3+} + 1/2H_2O$$

Fe^{3+} precipitates as Fe (OH)$_3$ or FeO(OH)$_2$. For the above reaction to proceed, sufficient O$_2$ should be present in the seawater. The most likely mechanism of increasing O$_2$ is photosynthesis by microorganisms. O$_2$ liberated by photosynthesis was consumed by the oxidation of Fe^{2+} in the early stages. After the formation of BIFs, the O$_2$ concentration of the atmosphere and oceans increased gradually. O$_2$ migrates in upward in the atmosphere. Ozone was formed by

$$O_2 + O \rightarrow O_3$$

Ozone absorbs ultraviolet radiation, leading to favorable conditions for the evolution of life.

Sources of Fe^{2+} in the oceans have been discussed. One of the likely sources delivering Fe^{2+} to the oceans is the interstitial water in seawater sediments containing high concentrations of Fe^{2+} (Holland 1973). There is a debate whether the mechanism whereby Fe^{2+} is oxidized is abiotic or biotic.

K-rich granitic rocks and polymetallic sulfide–sulfate ore deposits appeared between 3.0 and 2.5 Ga. In contrast, Archean granitic rocks are Na-rich. These changes relate to the beginning of plate subduction and associated igneous activity, the formation of island arcs and the growth of continents. Several continental growth

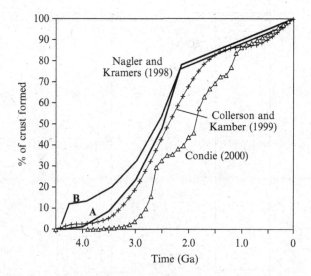

Fig. 6.5 Crustal growth curves based on transport-balance modeling (Nagler and Kramers 1998), U–Pb zircon ages from juvenile crust (Condie 2000) and Nb/U ratios for the depleted mantle (Collerson and Kamber 1999; Rollinson 2007)

rate models have been proposed (Fig. 6.5). We now think rapid growth of the continents from 3.0 to 2.5 Ga. seems plausible. The Nd/Sm isotopic ratio suggests step-wise continental growth rather than continuous, gradual growth. According to these processes, continental crust, oceanic crust, and mantle separated distinctly.

With the growing of continents, riverine input to the oceans increased. The continental rocks were weathered by surface water, and alkaline and alkaline elements were released from the rocks by the reaction

$$silicate + H^+ \rightarrow cation + clay\ mineral$$

The weathering rate is influenced by the CO_2 concentrations in the atmosphere and the surface water. High temperatures lead to higher weathering rates. We have concluded that riverwater supplied large amounts of Ca, Fe, Mn, Mg, and HCO_3^- to the oceans, and HCO_3^-, Ca, and Mg became fixed as carbonates there.

Marine biota plays an important role in the formation of carbonates as limestone in the oceans today. Thus, it is thought that older carbonates were also formed by marine biota with carbonate shells.

The last important environmental change is the increase in atmospheric O_2 caused by photosynthetic activity of organisms. Paleosols, redbeds, and mass-independent sulfur isotopic data indicate that atmospheric O_2 rose to appreciable levels about 2.3 billion years ago. The appearance of glaciers at 2.8 Ga also characterizes the surface environment of this period. This may relate to the decrease in atmospheric CO_2 concentration. It is also possible that the changes in continental area and distribution influenced climate change, leading to glaciation. In addition, planetesimals and meteorite impacts and igneous activities influenced the climate. The earth's interior changed considerably in this period. The growth of the inner core intensified the earth's magnetic field.

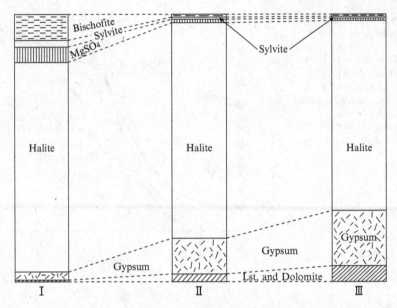

Fig. 6.6 Comparative precipitation profiles from I, the experimental evaporation of seawater, II, Zechstein evaporites; and III, the average of numerous other marine salt deposits (Borchert and Muir 1964; Holland 1984)

6.2.3   2.0 Billion Years Ago to Present

Numerous geological and geochemical data on this period, particularly the Phanerozoic age, are available.

The removal of O_2 by the formation of BIFs decreased after about 2.0 Ga, and the atmospheric O_2 concentration gradually increased from that time. This increase is estimated from the occurrence of red beds, sedimentary rocks with high Fe^{3+}/Fe^{2+}, around 1.6 Ga. Atmospheric CO_2 and the concentration of CO_2 in seawater decreased because of silicate weathering on the continents and carbonate precipitation in seawater. Most carbonate rocks which were formed related to biological activity in seawater appeared after about 1.9 Ga.

The concentrations of alkali ions, alkaline earth ions, SO_4^{2-}, Cl^-, and base metal elements in seawater have not varied significantly since that time. This idea is supported by evidence that the stratigraphic sequence of minerals in evaporites of this period is nearly the same then as now (Fig. 6.6). The ratio of Ca–Mg carbonate (dolomite) to Ca carbonate (calcite, aragonite) varies with age. This ratio is related to the ocean floor spreading rate. Increases in ocean floor spreading rate lead to increases in the seawater-hydrothermal solution cycling rate, causing the removal of Mg from cycling seawater, In contrast to the major element concentrations, isotopic compositions including the $^{87}Sr/^{86}Sr$ ratio, $\delta^{34}S$, $\delta^{13}C$, and $\delta^{18}O$ and global cycles of relative changes of sea level during Phanerozoic time have varied significantly

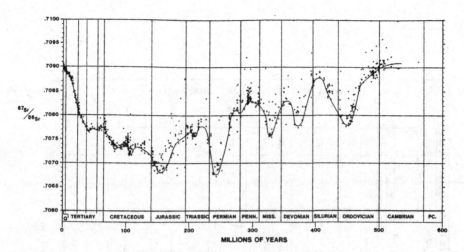

Fig. 6.7 First and second order global cycles of relative changes of sea level during Phanerozoic time (Vail et al. 1977; Holland 1984)

(Figs. 6.7 and 6.8). It is notable that the isotopic compositions covaried with each other. The low $^{87}Sr/^{86}Sr$ of seawater of this period is thought to be due to a high ocean floor spreading rate. When the spreading rate increased, the seawater–midoceanic ridge basalt interaction proceeded, resulting in decreased $^{87}Sr/^{86}Sr$ ratios in seawater. In general, the $^{87}Sr/^{86}Sr$ ratio of oceanic basalt is lower than that of seawater. This is due to the input of Sr derived from older continental by riverwater into the oceans. The $\delta^{34}S$ of hydrothermal seawater solutions originating at midoceanic ridges is low due to the contribution of sulfide sulfur in basalt. Therefore, hydrothermal solution inputs cause decreased $\delta^{34}S$ in seawater. If the above argument is correct, $\delta^{34}S$ and the $^{87}Sr/^{86}Sr$ ratio of seawater covary.

It is known that sea levels declined in this period, leading to enlargement of continental areas. In this situation, sulfate ions derived from continental sulfur with low $\delta^{34}S$ are transported by riverwater to the oceans, leading to decreased seawater $\delta^{34}S$. However, sulfate reduction by bacteria in seawater leads to increased seawater $\delta^{34}S$. If this process was intense, it becomes difficult to explain the covariation of the $^{87}Sr/^{86}Sr$ ratio and the $\delta^{34}S$ of seawater based on different supply rates from the continents.

Berner and Canfield (1989) obtained atmospheric O_2 levels from 5.7 hundred million years ago to the present using a mathematical calculation model. According to this calculation, atmospheric O_2 levels in the Carboniferous-Permian era (2.5–3.0 hundred million years ago) were high. The age of this oxygen catastrophe corresponds to high $\delta^{13}C$ and low $\delta^{34}S$ in seawater. Berner and Canfield (1989) thought that the sedimentation rate, or organic carbon and sulfide sulfur burial rate, is a more important factor in controlling the atmospheric O_2 level than weathering and basalt–seawater interaction. However, Holland (1984) explained high the $\delta^{13}C$ at that time by the influence of volcanic gas. In fact, in this age, huge basaltic eruptions caused by hot spots related to mantle plume activity occurred.

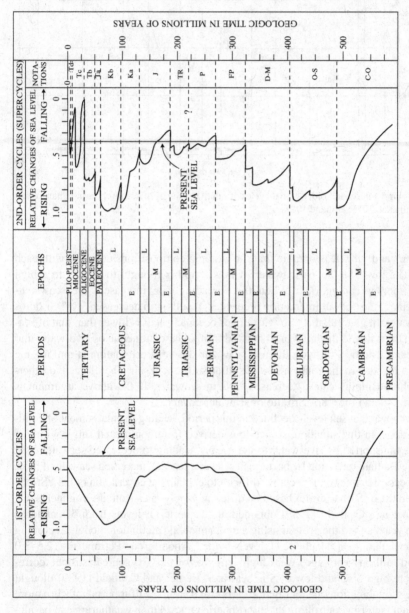

Fig. 6.8 Variation of the $^{87}Sr/^{86}Sr$ ratio of seawater during the Phanerozoic Eon (Burke et al. 1982; Holland 1984)

Fig. 6.9 Plots of worldwide mean annual atmospheric surface temperature versus time, based on BLAG model, compared with temperatures estimated from paleobotanical study and $\delta^{18}O$ of planktonic carbonate (35°N) (Berner et al. 1983)

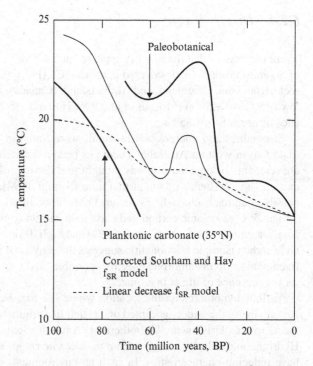

Berner and Canfield's model cannot explain the variation of the $^{87}Sr/^{86}Sr$ ratio. Therefore, it seems more likely that volcanic gas input and seawater cycling control the $^{87}Sr/^{86}Sr$ ratio and other isotopic compositions of oceans. The BLAG model (Berner et al. 1983) clearly indicated the relationship between plate tectonics and atmospheric CO_2 variation with time (Chap. 3). This model quantitatively interpreted the influence of continental area and the ocean floor spreading rate on atmospheric CO_2 and showed that the calculated temporal CO_2 variation is similar to results from paleobotany and oxygen isotopic compositions of planktonic shell carbonate (Fig. 6.9). This clearly indicates that tectonics, including the ocean floor spreading rate and continental area, which is also controlled by plate tectonics, is a very important factor controlling atmospheric CO_2 concentration and climate change. The BLAG model (Berner et al. 1983) estimated high atmospheric concentrations in the Cretaceous (65 million–100 million years ago) several times higher than the present level. This model simulates the global carbon cycle based on the reactions between carbonates, silicates, and CO_2 (e.g. $CaSiO_3 + CO_2 = SiO_2 + CaCO_3$). This implies that the CO_2 concentrations in the atmosphere and oceans are governed by these reactions. After the BLAG model was proposed, Berner (1994) proposed more convenient simplified models like the GEOCARB model and estimated Phanerozoic P_{CO_2} and P_{O_2}.

6.2.4   Origin of Life

There is evidence that the first life appeared at least by 3.8 Ga. Recently, fine grains of organic carbon were discovered in apatite ($Ca_5(PO_4)_3$ (OH,F)) in the sedimentary rocks from Isua, Greenland, and Akilia Island, (Canada), both from 3.85 Ga is very low (−50‰ to −20‰) (Mojzsis et al. 1996; Holland 1997). This suggests the presence of life before 3.85 Ga.

Stromalites and microorganism fossils were found in Warrawoona group rocks (3.45 Ga) in western Australia. There has been a debate whether these stromalites are remnants of organism. It was thought that the microfossil filaments were evidence of the existence of life at that time (Schopf 1994). However, it is also very possible that the bacteria-like structures are abiotic in origin.

The $\delta^{13}C$ of organic carbon and carbonate carbon from sedimentary rocks from Isua, Greenland (3.8 Ga) are low (−25‰) and high (0‰ to −3‰), respectively. This wide carbon isotopic fractionation suggests the activity of microorganism. However, fractionation by metamorphism is also possible, and thus it is not certain evidence for the presence of life at that time.

Various hypotheses on the location where life first began have been proposed. Among them, the widely accepted one is that life originated in the oceanic environment, particularly near hydrothermal solution vents at midoceanic ridges. Hydrothermal solutions issuing from midoceanic ridges contain CO_2 and CH_4 and have reducing characteristics. In such an environment, energy supplied from the earth's interior would be large enough to drive the inorganic reactions that promote chemical evolution and could lead to the generation of life.

Experiments have shown that cell-like microglobules and peptides were synthesized from hydrothermal solutions under high temperature and pressure (Yanagawa 1989). The other evidence supporting the "hydrothermal solution hypothesis" is the discovery of polyphosphoric acid from volcanic gas today. This polyphosphoric acid would have been an essential substance for synthesizing DNA (deoxyribonucleic acid) for the organisms before 4 Ga in the pre-biotic evolutionary stage. Hydrothermal solution issuing from midoceanic ridges contains considerable amounts of base metal elements (Fe, Mn, Zn, Cu, etc.). These elements in the solution and sulfides play roles as catalysts, promoting organic reactions. The "pyrite hypothesis" was proposed based on the idea that CO_2 is reduced to form various types of organic matter on the surface of pyrite crystals (Wachtershauer 1988). It was experimentally determined that the reaction of iron sulfide (FeS) changing to pyrite (FeS_2) generates H_2 by

$$FeS + H_2S \rightarrow FeS_2 + H_2$$

Thus, it may be possible that such H_2 reduces CO_2 to organic matter.

The universal tree of life derived by sequencing ribosomal RNA (Fig. 6.10) (Kump et al. 1999) shows that organisms can be divided into Bacteria, Archaea, and Eukarya. Bacteria and Archaea are composed entirely of single-celled organisms, while Eukarya includes higher plants and animals including humans. It is interesting to note that the point at which Archaea and Eukarya split from the Bacteria is

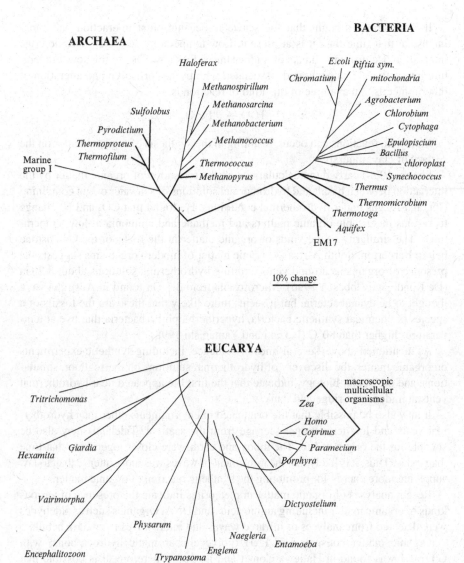

Fig. 6.10 The Universal Tree of Life devived by ribosomal RNA (Kump et al. 1999)

thought to lie close to the common ancestor of all life. In Fig. 6.10, the shaded branches near this point represent hyperthermophilic bacteria that live at temperatures above 80°C and are found at seafloor vent systems and terrestrial geothermal systems such as Yellowstone, USA. This may suggest that life originated at seafloor vent systems. It may be also likely that life originated in some cool surface environment and then proceeded to colonize most of the earth. Or, it is possible that a giant impactor hit the earth and destroyed all of the surface-dwelling organisms, but organisms from the vent systems could have begun the process of recolonizing (Kump et al. 1999).

It is generally thought that the seawater–oceanic crust interaction was more intense at that time than it is at present. Low temperature seawater–oceanic crust interaction occurs on the flanks of midoceanic ridges now. Due to this low temperature interaction, H_2 is generated, associated with pyrite formation and alteration of other minerals. An example of this kind of reaction is

$$2\text{"FeO"} + H_2O \rightarrow \text{"Fe}_2O_3\text{"} + H_2$$

Catalystic reactions can occur, resulting in the synthesis of organic matter on the surface of clay minerals.

Shock (1990) carried out calculations on the formation of organic matter by the interaction between basalt and hydrothermal solutions of seawater origin containing CO_2 and N_2 gas under hydrothermal conditions. He found that CO_2 and N_2 change to various metastable organic matters and methane and ammonia at low temperatures. The similarity of the kinds of organic matter in the rocks of the Greenstone belt in Barberton, South Africa to organic matter of modern organisms suggests the presence of organisms around vents emitting hydrothermal solutions about 3.2 Ga (De Ronde and Ebbesen 1996). Microfossils from 3.5 Ga found in Australia were thought to be cyanobacteria, but it seems more likely that these are the fossils of a species of chemical synthetic bacteria, hyperthermophilic bacteria that live at temperatures higher than 80°C (Isozaki and Yamagishi 1998)

As mentioned above, several lines of evidence, including synthetic experiments on organic matter, the discovery of hydrothermal solutions on the seafloor, simulations, and molecular biology, indicate that the first life appeared near hydrothermal vents at midoceanic ridge and flank.

It may also be possible that life originated in environments other than hydrothermal vents and low temperature seepage from the seafloor. Tidelands may also be suitable for the origin of life. The moon and earth were closer together at that time than today. Thus, it is likely that the movement of waves was more intense than today, supplying more energy for promoting the synthetic reaction of organic matter.

Recent analyses of organic matter in meteorites indicate the presence of various kinds of organic matter including amino acids and DNA. Organics such as aldehydes were detected from analyses of infrared waves and microwaves from dark nebula.

Organic matter consisting of PAH (polycyclic aromatic hydrocarbons) with C:O = 1:1 was found in Halley's Comet and its dust. Therefore, it is possible that extraterrestrial organic matter that originated in comets, particularly from the impact bombardment stage, could be the source of organic matter in life.

6.2.5 Extraterrestrial Life

In November 2000, a specific bacteria different from any of earth's bacteria was found in the atmosphere at an altitude of 16,000 m. Some scientists think this bacteria came from a comet. Thus it may be possible that life was extraterrestrial in origin.

The discovery of small carbonate grains similar in form to bacteria in a Martian meteorite underscored the possibility that extraterrestrial life may exist. However, recently inorganic origin of the carbonate is more accepted. $MgSO_4$, Na_2SO_4, and H_2SO_4 hydrate have been reported to exist on Ganymede. McCord et al. (1998) inferred that they formed through the interaction of sulfur and water and the evaporation of seawater rich in salts at the surface. It is thought that seawater exists below the ice crust. Seawater is not present on Titan, but N_2 is a major atmospheric gas. This is similar to the earth's atmosphere. It is said that methane exists in the atmosphere, there is an ethane ocean, and various kinds of organic matter may form on Titan's surface due to chemical reactions caused by meteorite impact, cosmic ray irradiation and thunder. It may be likely that life is present in Titan's seawater below the 1 km thick surface ice (Naganuma 2004). Traces of hydrothermal activity have been found on the Martian surface. There are many volcanoes that were active in recent times and calderas associated with them. Water ice is known to exist underground. Therefore, if magma intrudes into the ice, the ice melts, generating a hydrothermal solution. It may be possible that life originated under such condition. The $\delta^{34}S$ of sulfide in Martian meteorites (ACH84001, Nakhla) falls in a wide range from $-6.1‰$ to $+8.0‰$. Hydrothermal activity is thought to be the cause of this variation (Greenwood et al. 2000). Various lines of evidence show the existence of liquid water in the form of seawater, riverwater, and lakewater on the past Martian surface. The evidence includes features of the relief on the Martian surface, sediments formed in aquatic environments, the occurrence of hematite and jarosite, and the high water content of pyroxene in Martian rocks (Head et al. 1999). It is certain that magmatic activity has occurred. Therefore, if an ancient ocean existed, it may be likely that seawater cycling occurred and hydrothermal solutions issued from the seafloor. Life may have originated in such an environment, similar to the possibility on earth.

6.2.6   The Biosphere Deep Underground

Deep sea life living near hydrothermal vents takes chemical energy from hydrothermal solutions. This is different from photosynthetic biota. These biota (thernophite) are found near hydrothermal vents and seepage sites in subduction zones. Recently, similar life was found underground and the term "underground biosphere" was coined. This finding may suggest the existence of biota underground in France (L'Haridon et al. 1995). In 1996, a microorganism was found from an oil reservoir 4.2 km deep with a temperature of 110°C in Alaska (Stetter et al. 1993). The deepest that microorganisms have been found is 5.2 km where granitic rocks (Sweden) exist (Szevtzyk et al. 1994). There are reports of sulfate reducing bacteria in marine sediments 80 m below the sea floor under 900 m of water in the Sea of Japan. These are Desulvovibrio that can survive up to 65°C and 275 atm of pressure (Naganuma 2004).

6.2.7   The Evolution of Life

Carbon isotopic data on organic matter in sedimentary rocks from the Isua district, Greenland, are low, suggesting their organic carbon is of biogenic origin. Iron deposits and sulfide deposits are distributed in the metamorphic rocks of sedimentary origin from that area. Thus, it seems possible that life originated in the hydrothermal system associated with these ore deposits.

The oldest reported fossil is the 3.1-billion-year-old Figtree chert from South Africa. However, this was claimed to be a fossil extant at that time. It is bacteria-like in form and is named eobacterium. Stromalites have been discovered in rocks from Rhodesia, South Africa dated to 3.0 Ga. In these ages, O_2 was consumed by the oxidation of Fe^{2+} to Fe^{3+}, and thus no free O_2 was present in the oceans or atmosphere. After 2.3 Ga, atmospheric O_2 increased due to the increase in cyanobacteria, decreased oxidation of Fe^{2+} to Fe^{3+}, and a change in the proportion of reducing volcanic gases produced or to the way in which O_2 was utilized in oxidizing crystal rocks (Catling and Claire 2005; Rollinson 2007).

At the end of the Precambrian, the protozoa-like radiolaria and foraminifer appeared in the oceans. They were preserved as fossils that are different from previous biota. We can decipher the evolution of biota from these ages more clearly than from earlier ones.

We think that earth's entire surface environment then cooled and was covered by ice (the "snow ball earth"). The snow ball earth hypothesis was proposed based on evidence that: (1) BIFs are associated with glacier sediments, and (2) the $\delta^{13}C$ of carbonate rocks is $-5‰$, similar to mantle values, and does not show biological activity. The main cause for the cooling is uncertain, but it is thought to be a decrease in the rate of degassing of CO_2 from the earth's interior. By this cooling, the entire ocean was covered by ice, leading to the extinction of photosynthetic bacteria and other organisms since no sunlight passed through the surface. However, mass extinction did not occur. This may be due to global warming after this event. This warming could have been caused by increased degassing and methane generated by the melting of gas hydrates.

In the Paleozoic age (5.64–3.22 billion years ago), the diversity of species increased considerably. Various animals developed from the end of Proteozoic to the Cambrian, related to the ending of the glacial period at the late Proterozoic. Large amounts of nutrients dissolved into the seawater as the ice melted. Photosynthetic biota used nutrients in the seawater, and large amounts of O_2 were generated, leading to a rapid evolution of biota in the Cambrian. For example, Trilobites, a Paleozoic index fossil, appeared in the Cambrian and lived until the end of the Paleozoic. In the Silurian, plants appeared. In the Carboniferous-Permian, reptiles appeared and then developed in the Mesozoic. In the Mesozoic, ammonites lived in the oceans. Mammalians appeared in the Cenozoic and developed and evolved into the higher animals.

The evolution of life described above is shown in Fig. 6.11. As shown in this figure, the diversity of biota has increased over a very long time, and sometimes decreased by mass extinctions.

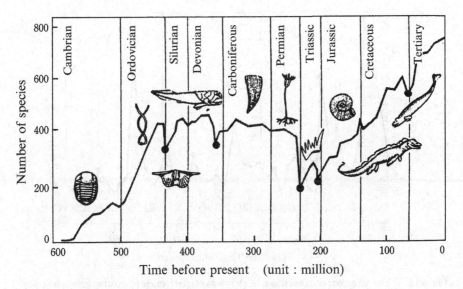

Fig. 6.11 Evolution of life (Shikazono 1995)

6.2.8   Mass Extinction

A thin clay layer enriched in iridium (Ir) and osmium (Os) was discovered at the Cretaceous/Tertiary (C/T or K/T) boundary (65 Ma) in sediments at Gubbio, Italy (Alvarez et al. 1980). This enrichment was later found in various localities in Italy, Denmark, New Zealand, and in New Mexico, USA. Ir and Os are very rare on the earth's surface. In contrast, meteorites contain high amounts of these elements. This anomalous enrichment at the K/T boundary was interpreted as being caused by the impact of a meteorite with a diameter of about 10 km moving 10 km/s. A giant explosion occurred with this impact, accompanied by the dispersion of fine solid particles and gases into the atmosphere, and a drop in temperature. After that, however, the temperature increased due to the greenhouse effect of water vapor. In the atmosphere, NOx compounds formed and dissolved into rain, resulting in acid rain. Such rapid climate change and acid rain caused mass extinctions. For example, the Age of Dinosaurs ended. Benthic reef communities were also extinguished. Other lines of evidence supporting the "meteorite impact hypothesis" include (1) shocked quartz grains occur characteristically in the K/T boundary layer. This texture is thought to have formed in a sudden high pressure event, (2) other siderophile elements that are rich in meteorites are found in the boundary layer, (3) the Os isotopic composition of the layer is similar to meteoritic values, (4) high $^{87}Sr/^{86}Sr$ values in sediments at this boundary are explained by contamination of terrestrial materials by the bolide impacting the earth, and (5) recently, one of the largest known craters (the Chicxulub crater), with a diameter of at least 200 km which formed at the age of boundary, was discovered in the area of the Yucatan Peninsula, Mexico. However,

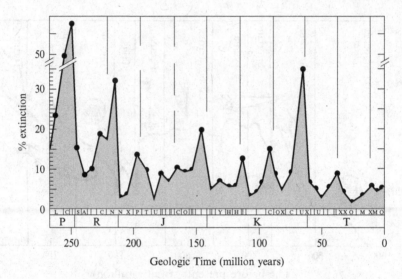

Fig. 6.12 The fossil record of extinction rate, shown as the percentage of existing genera that went extinct in a particular interval of geologic time. The vertical line indicates the 26 year periodicity of extinction (Sepkoski 1994)

another hypothesis that can explain the Ir enrichment has been proposed. Huge basaltic eruptions occurred on the Deccan Plateau of India at the time of the K/T boundary. This was one of earth's largest episodes of volcanic activity. Ir is detected in present-day volcanic gas such as on Kilauea in Hawaii. Therefore, it may be possible that the Ir enrichment was caused by the involvement of volcanic gas from the Deccan eruption. This hypothesis can also explain Os isotopic composition and minor elements contents.

Several other mass extinction events have occurred throughout the earth's history (Fig. 6.12). The largest one was at the Permian/Triassic (P/T) boundary. Almost 50% of marine organisms (ammonites, nautilus, etc.) went extinct during that event. The cause for this mass extinction is uncertain, although considerable research has been carried out. The possible causes are a decline of the sea level and decrease in the salinity of seawater by formation of evaporites and climate change. These are related to the formation of the Pangaea supercontinent as the continents collided. The formation of the supercontinent may have influenced climate change. As described before, at this boundary age, the $^{87}Sr/^{86}Sr$ ratio, $\delta^{34}S$, and $\delta^{13}C$ of seawater changed significantly. There is no clear direct evidence of meteorite impact at this boundary. Rapid eruption of the Siberians and the huge Traps flood basalt occurred at the P/T boundary (Renne and Basu 1991). It is possible that these eruptions caused the mass extinction (Kamo et al. 2003).

Raup and Sepkoski (1982) showed that eight major mass extinctions have occurred during the last 268 million years at intervals of about 26 million years (Fig. 6.13). Several hypotheses have been put forth to explain this periodicity in the fossil data of extinction rates. For example, scientists insisting on a bolide impact

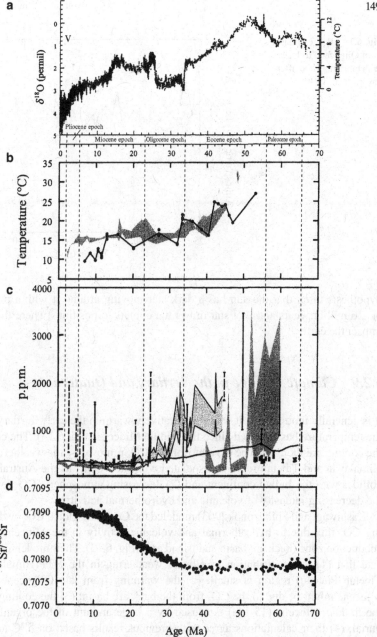

Fig. 6.13 Climate change interred from various climate proxies over the Cenozoic. (**a**) Oxygen isotope records (Zachos et al. 2003). (**b**) Terrestrial temperature estimated from the leaf-margin analysis (LMA) (*solid line*: Wolfe 1995) and the Coexistence Approach (CA) (*shaded area*: Mosbrugger et al. 2005). (**c**) Atmospheric carbon dioxide estimated from phytoplankton (*shaded area with a frame*: Pagani et al. 2005b), soil carbonate (*broken bars*: Ekart et al. 1999), Stomatal index (*solid bars*: Retallack 2001; Royer et al. 2001), boron isotope (*shaded area without a frame*: Demicco et al. 2003), and numerical modeling (*solid curve*: Kashiwagi et al. 2008). (**d**) Strontium isotope records (Veizer et al. 1999)

Fig. 6.14 P_{CO_2} estimates from Oligocene to late Miocene (modified after Pagani et al. 1999)

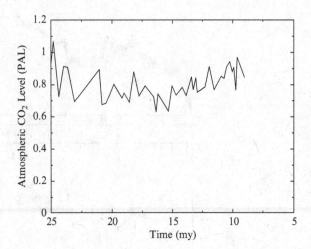

hypothesis think that the sun has a dark star moving around it with a periodicity of 26 million years and that star draws the comets to positions where their orbits impact the earth.

6.2.9   Climate Change in the Tertiary and Quaternary

It is generally thought that it was comparatively warm in the early Tertiary but that the temperature has declined since the middle Miocene (Fig. 6.13). The causes for the cooling are thought to be (1) atmospheric CO_2 removal by the weathering of the Himalayan and Tibetan regions associated with their uplift as the Australian plate collides with the Indian continent and (2) decreasing atmospheric CO_2 levels with the decreasing intensity of volcanic and hydrothermal activity.

Kashiwagi and Shikazono (2003) modeled the Cenozoic global CO_2 cycle including CO_2 flux due to hydrothermal and volcanic activity in the Miocene age at a subduction zone (back arc basin and island arc) (Fig. 6.13). Their model results indicate that (1) the contribution of silicate weathering in the HTP(Himalayan and Tibetan Plateau) region is small; (2) the warming from late Oligocene to early Miocene might be due to the CO_2 from the back arc basin; (3) the cooling event in the middle Miocene (15 Ma) is caused by a large amount of the organic carbon burial; (4) their calculations agree with previous results based on $\delta^{13}C$ and δB of foraminifer shell carbonates (Pagani et al. 1999; Pearson and Palmer 2000) (Figs. 6.14 and 6.15). This indicates that cooling occurred in the northern hemisphere, but the entire earth's surface temperature did not decline during the Miocene.

The Quaternary is characterized by repeated glacial–interglacial cycles causing wide environmental changes. Climate change and sea level variation are estimated from deep-sea sediment data. We can know the exact ages from microfossils like foraminifera and radiolaria in sediments, paleomagnetic data, radiogenic ages and

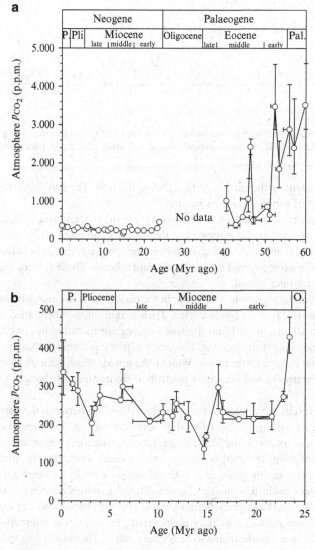

Fig. 6.15 Atmospheric CO_2 data for the past 60 m.y. (**a**) The entire record and (**b**) an enlargement of the past 25 m.y. (modified after Pearson and Palmer 2000)

sedimentation rates. The isotopic composition of oxygen in seawater can be estimated from that of foraminiferal carbonate shells (Fig. 6.16). High $\delta^{18}O$ values in seawater means the enlargement of continental ice areas that occurs in cold climates. The assemblage of radiolaria is useful in determining surface seawater temperatures. Oxygen isotopic studies indicate that eight glacial–interglacial cycles have occurred during the last 7×10^5 years (Fig. 6.16). From the changes in the orbital parameters eccentricity, tilt and precession, we can calculate the sunlight

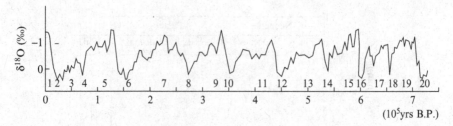

Fig. 6.16 Variation of seawater $\delta^{18}O$ since 7×10^5 years ago (Sugimura et al. 1988). 1–20: Interglacial–glacial stage number. Odd and even number is glacial and interglacial stage, respectively

radiation incident on the earth's surface during the past. The calculated results agree with the $\delta^{18}O$ of foraminiferal carbonates.

Variation of atmospheric CO_2 concentration in the past can be estimated from ice core data. Ice core analyses covering the last 15 thousand years detail climate changes during that period. Other factors influencing interglacial cycles are ocean circulation, the area covered by ice sheets and albedo. These factors are not independent but are interrelated.

Shorter cyclical variations also occur in the earth's surface environment. Sunspots wax and wane in 11- and 22-year cycles. Their maximum number varies in a 50-year cycle. Scientists discovered that the sun's cycles are recorded in climate data like regional floods and temperatures. The sun's activity influences short-term climate changes in the earth's surface environment. As noted above, various short-term and long-term, externally and internally controlled environmental changes occur in the earth system.

External forces are related to the spatial relationship between the earth and sun, sun activity, and the spatial relationship between the sun and the galaxy's spiral arms. External forces come from plate motion, mantle convection and mantle plumes approaching the earth's surface environment. For example, mantle plume activity has been taking place in a 200-million-year cycle for large activity and a 30-million-year cycle for smaller plumes. These activities accelerate mantle convection and ocean floor spreading and cause climate change and the extinction of biota. The cause of this periodicity is uncertain. However, the interaction between the core and lower mantle seems to be a likely cause. The hypothesis that meteorite impacts cause mantle plume activity periodicity also seems likely. If this hypothesis is correct, we cannot understand short- or long-term changes in the earth's surface environment without taking into account both external and internal influences.

6.2.10 Anthropogenic Influence on the Earth's Changing Environment

Anthropogenic influence on the earth's environment before 8,000 years ago, when agriculture began, was negligible. However, after the start of agriculture global warming began. Previously it was thought that greenhouse gases emitted by human

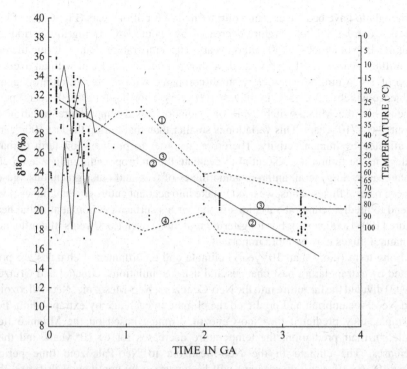

Fig. 6.17 Compilation of $\delta^{18}O$ variation of cherts with time (Knauth and Lowe 1978; modified after Holland 1984). *Curve* ① connects the largest δvalues for cherts at any given time. This is the current best estimate of the oxygen isotopic composition of cherts formed in marine or near-marine environments. *Curve* ② represents the initial estimate of this secular change by Perry (1967). *Curve* ③ is the estimate by Perry et al. (1978). *Curve* ④ connects the lowest δ values reported for cherts at any given time

activity had influenced global warming only during the last 200 years (Sect. 6.2.9). However, Ruddiman (2003) analyzed variations of CO_2 concentration obtained from ice cores from Vostok in the Antarctic, and recognized that the atmospheric CO_2 concentration during the last 8,000 years deviated from the natural background trend. In the late stone age, 8,000 years ago, deforestation for the cultivation of wheat, barley, and peas started in Europe. The atmospheric methane concentration increased suddenly. This may be due to the beginning of rice cropping 5,000 years ago in southern China.

6.2.11   Prediction of Earth's Environment in the Future

Can we predict the earth's environment in the future based on estimated variations of its environment in the past and knowledge of the earth's environment today? First let us consider climate change. As shown in Fig. 6.17, the temperature of seawater is thought to have been decreasing during the last 3 billion years (Fig. 6.17). If this curve is correct, the temperature decreased by about 60°C during this period, an average drop of about 1°C/50 million years. The temperature change during the last 65 million years is well investigated. It shows a decreasing trend with an average drop of 1°C/4 million years, very small compared with the recent anthropogenic influence which is 3 ± 1.5°C in the last 100 years. Short-term (10^4–10^5 years) periodicity like the Milankovitch cycle of glaciation has a temperature variation of about 3–6°C/10^5 years. This variation is smaller than that during the last 100 years, as affected by human activity. Therefore, in order to predict short-term climate change in the future, it is essential to evaluate the anthropogenic influence on climate change. In general, anthropogenic fluxes of CO_2 and other greenhouse gases exceed natural fluxes. Thus, we need to take into account anthropogenic fluxes if we intend to predict temporal variations in the earth's surface environment in the near future (100–1,000 years). For long-term predictions over 1,000 years, the influence of natural fluxes may be more important.

Long-term (more than 10^4 years) climate and environmental changes are predicted by extrapolating past changes and using simulations. Graedel and Crutzen (1993) divided the far future into the Neo-Cenozoic, Neo-Mesozoic, Neo-Paleozoic, and Neo-Precambrian and predicted the climate in each age by extrapolation. For example, they predicted the Neo-Cenozoic climate based on the Milankovitch cycle. In their prediction, the temperature decreases for 6×10^4 years and then increases. The climate in the Neo-Mesozoic to Neo-Paleozoic time period (6.5×10^6–6×10^7 years from now) will be governed by continental drift and the distribution of the continents. At 1.5 billion years, most of the continents are distributed near the equator, resulting in a smaller area covered by ice in polar regions but an increase in the absorption of the sun's radiant heat. Due to the decrease in albedo, the average surface temperature increases. The sun luminosity increases to 1.3–1.6 times the present in the Neo-Precambrian. Due to this effect, the rate of continental weathering increases and the atmospheric CO_2 concentration decreases, resulting in the extinction of plants. Even if atmospheric CO_2 decreases, the surface temperature may reach 100°C due to the sun's increasing luminosity after 1.3 billion years. According to Caldeira and Kasting's simulation (Caldeira and Kasting 1992), photosynthetic plants disappear after 10 billion years, organisms except bacteria are extinct after 1.3 billion years and no organism can survive after 1.5–2.5 billion years.

Studies on the environmental changes coming in the future have just started. In order to perform future predictions, it is essential to understand the various temporal scales and periodicities of the earth's environment and the mechanisms of those periodicities. The earth's surface reservoir is influenced not only by the other reservoirs on the surface, but also by invasion of matter and heat from earth's interior and from extraterrestrial bodies that sometime cause mass extinction. We have to recognize that the earth's surface system is open to its interior system and to the universal system regarding energy and matter in order to understand these systems.

6.3 Chapter Summary

1. The earth formed 4.6–4.5 billion years ago by the accumulation of planetesimals.
2. The primary atmosphere formed by the evaporation of volatiles in planetesimals and the involvement of primordial gas in the protosolar system. The primary atmosphere captured by protoearth was released during its T-Tauri stage. After that, a secondary atmosphere formed through degassing from the earth's interior.
3. Degassing occurred very rapidly in the early stage of the formation of the earth and did not occur continuously and gradually through Earth's history.
4. There is considerable debate whether the early-middle (until about 2.2 billion years ago) surface of the earth, in particular the atmosphere and oceans, was reducing or oxidizing. However, it is certain from much geochemical data that free oxygen was not present.
5. The planets in the solar system except the earth that were covered by CO_2 or liquid water are Mars and Venus. Comparisons of the earth with these planets (comparative planetology) can provide constraints on the origin and evolution of the earth.
6. The oldest rocks that are widely distributed on the earth is 3.8 Ga. Thus, we can discuss the evolution of the atmosphere, oceans, crust, and mantle during the last 3.8 Ga based on geological data such as ore deposits, sedimentary rocks, igneous rocks, and metamorphic rocks and geochemical data like the chemical compositions of major and minor elements, and isotopic compositions. Comparative planetology and computer simulations can constrain the environmental conditions before 3.8 Ga.
7. It is estimated that the earth's environment, in particular the oceans, atmosphere, and crust changed significantly during the period from 3.0 to 2.0 billion years ago. These changes were caused both by internal influences including plate tectonics, plume tectonics, and inner core-outer core interactions and an external one, the termination of planetesimal impacts.
8. Variations in the chemical compositions and isotopic compositions ($^{87}Sr/^{86}Sr$, $\delta^{34}S$, $\delta^{18}O$, $\delta^{13}C$, etc.) of seawater were determined by the cycling of seawater and hydrothermal solutions at midoceanic ridges, sedimentation and precipitation to the seafloor, transport of sediments by riverwater and glaciers and biological activity.
9. It was warm during the Archean eon. From the late Cretaceous to the present, the surface temperature has been decreasing as a general trend. Quaternary climate change is well investigated and a good correlation between atmospheric CO_2 concentration and surface temperature has been found. The Milankovitch cycle characterizes Quaternary environmental changes.
10. Various changes in earth's surface environment (atmospheric temperature, CO_2, isotopic composition, chemical composition, extinction of biota, igneous activity, distribution of continents, and tectonics) are interrelated. There are internal

factors (mantle plume activity and plate tectonics) and external factors (the spatial relationships between the solar system and galaxy, meteorites, and planetesimal impacts) causing these variations.

11. The earth's surface environmental system is open to the earth's interior system and the universe system with respect to energy and matter.

12. Anthropogenic and natural influences are necessary to evaluate short-term and long-term predictions of earth's environment in the future. Particularly, elucidation of various short-term and long-term periodicities and mechanisms causing environmental variations are necessary.

References

Alvarez CW, Alvarez W, Asaro F, Michel HV (1980) Extraterrestrial cause for the Cretaceous–Tertiary extinction. Science 208:1095–1108

Berner RA (1994) GEOCARB II: A revised model of atmospheric CO_2 over Phanerozoic time. Am J Sci 294:56–91

Berner RA, Canfield DE (1989) A new model for atmospheric oxygen over Phanerozoic time. Am J Sci 289:333–361

Berner RA, Lasaga AC, Garrels RM (1983) The carbonate-silicate geochemical cycle and its effects on atmospheric carbon dioxide over to past 100 milliion years. Am J Sci 283:641–683

Borchert T, Muir RO (1964) Salt deposits. Their origin metamorphism, and deformation of evaporites. Van Nostrand, New York

Brown H (1949) Rare gases and the formation of the Earth's atmosphere. In: Kuiber GP (ed) Atmosphere of the Earth and planets. University of Chicago Press, Chicago, pp 258–266

Burke WH, Denison RE, Hetherington EA, Koepnick RB, Nelson HF, Otto JB (1982) Variation of seawater $^{87}Sr/^{86}Sr$ throughout Phanerozoic time. Geology 10:516–519

Caldeira K, Kasting JF (1992) The life span of the biosphere revisited. Nature 360:721–723

Catling DC, Claire MW (2005) How Earth's atmosphere evolved to an oxic state: a status report. Earth Planet Sci Lett 237:1–20

Collersun KD, Kamber B (1999) Evolution of the continents and the atmosphere inferred from Th-U-Nb systematics of the depleted mantle. Science 283:1519–1522

Condie KC (2000) Episodic continental growth models: after thoughts and extensions. Tectonophysics 322:153–162

De Ronde CEJ, Ebbesen TW (1996) 3.2 billion years of organic compound formation near seafloor hot springs. Geology 9:791–794

Demicco RV, Lowenstein TK, Hardie LA (2003) Atmospheric pCO_2 since 60 Ma from records of seawater pH, calcium and primary carbonate mineralogy. Geology 31:793–796

Ekart DD, Cerling TE, Montarez IP, Tabor NJ (1999) A 400 million year carbon isotope record of pedogenic carbonate: implications for paleoatmospheric carbon dioxide. Am J Sci 299:805–827

Graedel TE, Crutzen PJ (1993) Atmospheric change—an Earth system perspective. W. H. Freeman, New York

Greenwood JP, Mojzisis SJ, Coath CD (2000) Surfur isotopic compositions of individual sulfides in Martian meteorites ALH 84001 and Nakhla: implications for crust-regolith exchange in Mars. Earth Planet Sci Lett 184:23–35

Hamada T (1986) Invitation to Earth science. University of Tokyo Press, Tokyo (in Japanese)

Head JW, Hiesimger H, Ivanov MA, Kreslavsky MA, Pratt S, Thomson BJ (1999) Possible ancient oceans on Mars: evidence from Mars Orbiter Laser Altimeter Data. Science 286:2134–2137

Holland HD (1972) The geologic history of seawater—an attempt to solve the problem. Geochim Cosmochim Acta 36:637–651

Holland HD (1973) The oceans: a possible source of iron in iron formations. Econ Geol 68:1169–1172

Holland HD (1984) The chemical evolution of the atmosphere and oceans. Wiley, New York

Holland HD (1997) Evidence for life on earth more than 3,850 million years ago. Science 275:38–39

Kamo SC, Czamanske GK, Amelin Y, Fedorenko VA, Davis DW, Trofimov VR (2003) Rapid eruption of Siberian flood-volcanic rocks and evidence for coincidence with the Permian-Triassic boundary and mass extinct at 250 Ma. Earth Planet Sci Lett 214:75–91

Kashiwagi H, Shikazono N (2003) Climate change during cenozooic inferred from global carbon cycle model including igneous and hydrothermal activities. Palaeogeogr Palaeoclim Palaeoecol 199:167–185

Knauth LP, Lowe DR (1978) Oxygeñ isotope geochemistry of cherts from the Onverwach Group (3.4 billion years), Transvacl, South Africa, with implications for secular variations in the isotopic composition of cherts. Earth Planet Sci Lett 41:209–222

Kump LR, Kasting JR, Crane RG (1999) The Earth system. Pearson Prentice Hall, Upper Saddle River

L'Haridon S, Reysenbacht AL, Glenat P, Prieur D, Jeanthan C (1995) Hot subterranean biosphere in a continental oil reservoir. Nature 377:223–224

McCord TB, Hansen GB, Fanale FP, Carlson RW, Matson DC, Jonson TV, Smythe WD, Crowley JK, Martin PD, Ocampo A, Hibbitts CA, Granhan JC (1998) Salton Europa's surface detected by galileo's near infrared mapping spectrometer. Science 280:1242, the NIMS Team

Mojzsis SJ, Arrhenius G, McKeegan KD, Harrison TM, Nutman AP, Friend CRL (1996) Evidence for life on Earth before 3800 million years ago. Nature (London) 384:55–59

Mosbrugger V, Utescher T, Dilcher DL (2005) Cenozoic continental climatic evolution of Central Europe. Proc Natl Acad Sci 102:14964–14969

Naganuma T (2004) Europa, life star. NHK Books, Tokyo (in Japanese)

Nagler TF, Kramers JD (1998) Nd isotopic evolution of the upper mantle during the Precambrian: models data and the uncertainty of both. Precamb Res 91:233–252

Ojima M (ed) (1990) Introduction to geophysics. University of Tokyo Press, Tokyo (in Japanese)

Pagani PN, Freeman KH, Arthur MK (1999) Late Miocene atmospheric CO_2 concentrations and the expansion of C4 grasses. Science 285:876–879

Pearson PN, Palmer MR (2000) Atmospheric carbon dioxide concentrations over the past 60 million years. Nature 406:695–699

Press F, Sieber R (1994) Earth. W. H. Freeman, New York

Raup DM, Sepkoski JJ (1982) Mass extinctions in the marine fossil record. Science 231:833–836

Renne PR, Basu AR (1991) Rapid eruption of the Siberian Traps flood basalts at the Permo-Triassic boundary. Science 253:176–179

Retallack G (2001) Cenozoic expansion of grasslands and climatic cooling. J Geol 109:407–426

Ringwood AE (1979) Origin of the Earth and Moon. Springer, New York

Rollinson H (2007) Early Earth systems. Blackwell, Oxford

Royer DL, Wing SL, Beerling DJ, Jolley DW, Koch PL, Hickey LJ, Berner RA (2001) Paleobotanical evidence for near-present-day levels of atmospheric CO_2 during past of the Tertiary. Science 292:2310–2313

Rubey WW (1951) Geologic history of seawater—An attempt to state the problem. Geol Soc Am Bull 62:1111–1148

Ruddiman WF (2003) The anthropogenic greenhouse era begun thousands of years ago. Clim Chang 61(3):261–293

Schopf JW (1994) The oldest known records of life: early Archean stromatolites, microfossils and organic matter. In: Begtson S (ed) Early life on Earth: Nobel Symposium, 84, pp 193–206

Sepkoski JJ (1994) Extinction and the fossil record. Geotimes 39:15–17

Shikazono N (1995) New Geology. Keiotsushin (in Japanese)

Shikazono N (1997) Carbon dioxide partial pressure of ancient earth's atmosphere deduced from carbonate-silicate eauiliblia. Yusenjin 6:186–195 (in Japanese)

Shikazono N (2010) Environmental geochemistry of Earth system. University of Tokyo Press, Tokyo (in Japanese)

Shock (1990) Do amino acids equilibrate in hydrothermal fluids? Geochim Cosmochim Acta 54:1185–1189

Siever R, Grotzinger J, Jordan TH (2003) Understanding Earth. W. H. Freeman, New York

Stanley SM (1999) Earth system history. W. H. Freeman, New York

Stetter KO, Huber R, Bochl EB, Kurr M, Eden RD, Fielder M, Cssh H, Vance I (1993) Hyperthermophilic archaen are thriving in deep North sea and Araskan oil reservoirs. Nature 365:743–745

Sugimura A, Nakamura Y, Ida Y (eds) (1988) Illustrated Earth science. Iwanami Shoten, Tokyo (in Japanese)

Szevtzyk U et al (1994) Thermophilic anaerobic bacteria isolated from a deep borehole in granite in Sweden. Proc Natl Acad Sci USA 91:1810–1813

Vail PR, Mitchum RM Jr, Thompson S III (1977) Seismic stratigraphy and global changes of sea level, Part 4, Global cycles of relative changes of sea level. In: Payton CE (ed) Seismic strarigraphy: application to hydrocarbon exploration, vol 26. American Association of Petroleum Geologists Memoir, Tulsa, pp 83–87

Veizer J (1976) In: Windley BF (ed) The early history of the Earth. Wiley, New York, pp 569–578

Veizer J, Ala P, Azmy K, Bruckschen P, Buhl D, Bruhn F, Carden GAF, Dienner A, Ebneth S, Podlaha OG, Strauss H (1999) $^{87}Sr/^{86}Sr$, $\delta^{13}C$, and $\delta^{18}D$ evolution of Phanerozoic seawater. Chem Geol 161:59–88

Wachtershauer G (1988) Before enzymes and templates: theory of surface metabolism. Microbiol Rev 52:452–482

Wolfe JA (1995) Paleobotanical interpretation of Tertiary climates in the Northern Hemisphere. Am Sci 66:694–703

Yanagawa H (1989) Investigation of origin of life. Iwanami Shoten, Tokyo (in Japanese)

Zachos JC, Wara MW, Bohaty S, Delaney MC, Petrizzo MR, Brill A, Bralower TJ, Premolisilva F (2003) A transient rise in temperature during the Paleocene–Eocene thermal maximum. Science 320:1551–1554

Chapter 7
A Modern View on Nature and Humans

As described in previous chapters, large amounts of information on spatial and temporal variations in the earth system are available. Therefore, people today can view nature on a global scale. They have information about the earth and can envision it as a system. People's images of the earth system vary by individual person, region, and country. It has been pointed out below that there are various views of the earth system, of humans and the relationship between the two. Here we would like to clarify the difference between the view in this book and others previously proposed. Modern views of the earth system and humans could be divided into human-oriented views (anthropocentrism) and nature-oriented ones (naturecentrism). Recently, a view unifying these two views has been proposed.

Keywords Anthropocentrism • Earth and planetary system sciences • Earth's environmental co-oriented society • Gaia • Naturecentrism

7.1 Anthropocentrism

Anthropocentrism considers the following points:

1. Humans have taken natural resources and used them to develop industry since the industrial revolution. Industry and economics developed significantly through this use of resources, but a large amount of waste was abandoned, considered outside the system of human society, resulting in earth's environmental pollution.
2. There is a view that considers resource and environmental problems as a whole. This is an extension of point 1. At present, various countries intend to develop new societies that can grow sustainably based on new scientific and information technologies.
3. It is difficult to solve the waste and environmental problems based only on points 1 and 2, and so the concepts of recycling (circulation) and the zero emission-type

N. Shikazono, *Introduction to Earth and Planetary System Science*,
DOI 10.1007/978-4-431-54058-8_7, © Springer 2012

society have been proposed recently. These are based on the concepts of circulation and symbiosis in natural systems, and recycling in human society. The zero emission concept was proposed at the World Environmental Conference in Rio de Janeiro, Brazil in 1992 sponsored by the United Nations. It is defined as "no emission to the hydrosphere and atmosphere and transformation of all waste to source materials".

Views 1, 2, and 3 regard nature and the earth as useful materials and energy available for human to use.

7.2   Naturecentrism

Naturecentrism encompasses the following four points:

1. There is a view that the earth consists of the atmosphere, hydrosphere, biosphere, and geosphere, but not humans. This is based on material and analytical sciences and reductionism. Previous earth and planetary sciences have been based on this view. Earth science is divided into geology, geophysics, geochemistry, geography, and oceanography, among others, each having been developed independently.
2. Recently, solid earth science was unified by the combination of plate tectonics and plume tectonics. Plate tectonics and plume tectonics are a dynamic view of the solid earth, while previous solid earth sciences were static views. Plate tectonics and plume tectonics consider the dynamics of plate motion. Geological and geophysical phenomena are caused by plate tectonics, but its principles do not include considerations of mass transfer mechanisms, nature–human interactions, or fluid earth-solid earth interactions.
3. Gaia (the mother earth goddess) hypothesis proposed by James Lovelock and Lynn Margulis regards earth as an ecosystem (Lovelock 1979, 1991). This hypothesis proposes that the earth can be viewed as a living organism that acts to manipulate the chemical and physical environment for the benefit and maintenance of life on earth. According to this view, humans are a part of the ecosystem and considering issues between humans and the earth is not necessary. This is a nature-, not human-oriented view. Its hypothesis is that the physicochemical environments of the earth, the atmosphere, and the oceans are controlled by the biota. James Lovelock indicated that the atmospheric composition of the earth is considerably different from that of Venus and Mars, and draws the conclusion that the earth's atmospheric composition is controlled by its biota. He thinks that the earth has been a self-regulating system, as demonstrated by his Daisyworld climate simulation, since the beginning of biological life. Atmospheric CO_2 eventually decreased due to the precipitation of carbonates from the oceans. Earth's surface temperature had been high, insulated by the greenhouse effect that atmospheric CO_2 causes. Plates that had high temperatures because of the greenhouse effect did not undergo subduction. However, the decrease in

atmospheric CO_2 caused by biological activity resulted in the cooling of plates and the beginning of plate motion. James Lovelock favors the idea that biological activity influences the dynamics of the earth's interior. In contrast, most earth scientists infer that long-term atmospheric CO_2 variation is governed by inorganic carbonate-silicate reactions as simulated in the BLAG model by Berner et al. (1983). It is certain that biological activity is important at least for short-term atmospheric CO_2 variations.

4. Previously earth system science considered the earth system as composed of subsystems and tried to elucidate the interactions between those subsystems and temporal variations of the earth system as a whole. The main object of previous earth system science was earth's surface system (the atmosphere, hydrosphere, and biosphere) but the solid bulk of the earth and the humans living on the surface were not the main objects. Earth system science as presented here aims to unify the dual concepts of anthropocentrism and naturecentrism, and others paired matters such as material science and history, resources and the environment, humans and nature, the fluid earth and the solid earth, parts and the whole, diversity and uniformity, accident and necessity, internal and external forces, and so on.

In addition to the views mentioned above, there are various views of nature and humans such as ecology, environmentalism, and geology. Traditional geology, which focuses on geologic history and description of the earth's materials and phenomena (volcanism, hydrothermal activity, sedimentation, metamorphism, folding, faulting, mountain building, etc.), particularly rocks and minerals, was developed based on uniformitarianism. Recently, however, catastrophic events have come to light. One example of this is the bolide impact hypothesis based on Ir enrichment at the K/T boundary (Alvarez et al. 1980). Dynamic view of the earth embodied by plate tectonics and the discovery of numerous craters on the moon and planets have caused the revival of catastrophism.

Uniformitarinism and catastrophism as described above are views of the earth that consider temporal but not spatial variations. A view of the earth and planets integrating spatial and temporal variations should be developed.

7.3 A New View of Nature and Humans

7.3.1 Earth Environment Co-oriented Society

Figure 7.1 shows human society as a system open to the natural system with respect to materials and energy. Materials and energy circulate in human society. They are input to human society and output from it to the natural system. Previous investigations have studied various environments, the atmosphere, hydrosphere, and biosphere in detail, but studies of the geosphere from an environmental point of view are lacking.

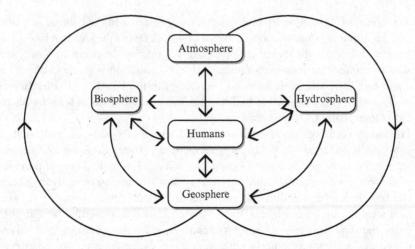

Fig. 7.1 Human society which is open to the natural system concerning material and energy

For example, we have plans to store and dispose of CO_2 and high-level nuclear waste deep underground. Therefore, humans must come to understand the geosphere as well as the atmosphere, the hydrosphere, and the biosphere, as well as the interactions between the geosphere and the other subsystems and learn to coexist with these subsystems. We describe this improved human society as an "earth environment co-oriented society".

7.3.2   Earth and Planetary System Science

Humans take natural resources, use them and dispose of the wastes they generate. The wastes emitted by humans influence the natural system. The natural system, changed by those wastes, influences humans and the biosphere. This negative feedback causes our environmental and resource problems. Materials and energy flow in the human–nature system, and interactions between subsystems can be investigated by earth system sciences. The earth system is open to the universe system. Earth and planetary system science intends to understand the earth system, the planetary system and the relationship between them.

As mentioned above, there are various views of the earth, planets, nature, and humans. Therefore, we have to make our viewpoint clear. It is important to recognize that humans are a part of the earth and planetary system, and further, of the system of the entire universe. We need to quantitatively describe the complicated hierarchy of the system, deduce its temporal and spatial variations, and analyze the interactions in the nature–human system. Previous earth sciences focused on the history of the earth but did not analyze the current earth, including the humans living on it. We need to predict not only short-term but also long-term environmental variations in the future, together with analyzing earth's former and present-day environments.

The important point in predicting the environmental conditions in the future is to assess simulation results. We can assess short-term future environmental variations, but unfortunately we cannot assess long-term ones.

7.4   Chapter Summary

1. We have various views on nature and humans. They are divided into human-oriented (anthropocentrism) and nature-oriented (naturecentrism) views.
2. Humans are a part of the earth, as well as the planetary and universal systems.
3. The concept of an "earth environment co-oriented society" is proposed here.
4. Earth and planetary system science has the potential to unify the anthropocentric and naturecentric views of the earth and planets.

References

Alvarez LW, Alvarez W, Asaro F, Michel HV (1980) Extra terrestrial cause for the cretaceous-tertiary extinction. Science 208:1095–1108

Berner RA, Lasaga AC, Garrels RM (1983) The carbonate-silicate geochemical cycle and its effects on atmospheric carbon dioxide over to past 100 million years. Am J Sci 283:641–683

Lovelock JE (1979) Gaia: a new look at life on Earth. Oxford University Press, Oxford, 157 pp

Lovelock JE (1991) Healing Gaia: practical science of planetary medicine. Oxford University Press, Oxford, 192 pp

Afterword

As already mentioned, this book focuses on (1) the constituent materials of the earth and planets (minerals, rocks, water, carbon dioxide, ore deposits, soils, atmosphere, etc.); (2) interactions between subsystems (atmosphere, hydrosphere, geosphere, biosphere, humans), particularly near the earth's surface environment; (3) dynamics of the earth system (plate tectonics, plume tectonics, global geochemical cycles including the earth's interior); (4) nature–human interaction (disasters, resources, environmental problems); (5) origin and evolution of the earth and planetary system; and finally, (6) the proposal that earth and planetary system science provides the scientific basis for considering the relationship between nature and humans and integrates previous disciplines of earth and planetary sciences (geology, geochemistry, geophysics, geography, etc.). It also proposes that earth and planetary system science gives us the basis of the concept of "earth environmental co-oriented human society," which could be safe, stable, and sustainable in the long term.

Earth and planetary system science treats entire fields of traditional earth and planetary sciences (e.g., geology, geochemistry, geophysics, geography, comparative planetology). However, it is very difficult to integrate them. Some fields are not described enough in this book. They include biological systems, planets and the universe, the processes of very short temporal scale and small scale (e.g., nano, atomic scale), predictions of the earth's future environment, and chemical and physical mechanisms of elemental and isotopic migration in the earth and planetary system. Among those, the chemical and physical mechanisms of elemental migration in the earth system are given in *Chemistry of the Earth System: Prediction of Environment and Resources* (1997) and *Environmental Geochemistry of the Earth System* (2010), which I wrote in Japanese. I am preparing the English version of these books and am hopeful that they could be published in the near future. I hope that earth and planetary system science including all of the fields mentioned above can be systematically integrated and synthesized in the near future.

Finally, for readers who want to study and understand more deeply earth and planetary system science and each field of earth and planetary sciences, I would like to recommend the following outstanding books.

N. Shikazono, *Introduction to Earth and Planetary System Science*,
DOI 10.1007/978-4-431-54058-8, © Springer 2012

Earth and Planetary System Science and the Global Geochemical Cycle

Berner RA (2004) The phanerozoic carbon cycle. Oxford University Press
Chameides WC, Perdue EM (1997) Biogeochemical cycles. Oxford University Press
Ehlers E, Kraft T (eds) (2006) Earth system science in the anthropocene. Springer
Ernst WG (ed) (2000) Earth systems. Cambridge University Press
Kump LR, Kasting JR, Crane RG (1999) The Earth system. Pearson-Prentice Hall
NASA (1986) Earth system science – overview – a Program for global change. Earth System Sciences Committee, NASA Advisory Council, Washington, DC
Shikazono N (1992) Introduction to Earth system. University of Tokyo Press (in Japanese)
Shikazono N (2009) Introduction to Earth and planetary system science. University of Tokyo Press (in Japanese)
Rollinson H (2000) Early Earth system. Blackwell

Earth's Environment and Resources

Craig JR, Vaughan DJ, Skinner BJ (1988) Resources of the Earth. Prentice Hall. Englewood Cliffs, NJ
Golob R, Brus E (1993) The Almanac of renewable energy. Holt, New York
Holland HD, Petersen U (1995) Living dangerously. Princeton University Press
Marini L (2007) Geological Sequestration of carbon dioxide. Elsevier
Shikazono N (2003) Geochemical and tectonic evolution of arc-backarc hydrothermal systems: implication for the origin of Kuroko and epithermal vein-type mineralizations and the global geochemical cycle. Developments in Geochemistry 8, Elsevier
Skinner BJ (1976) Earth resources, 2nd edn. Prentice Hall.

Evolution of Earth

Bengston S (ed) (1994) Early life on Earth. Columbia University Press, New York
Cloud P (1988) Oasis in space: Earth history from the beginning. Norton, New York
Frakes L A (1979) Climates throughout geologic time. Elsevier, Amsterdam
Holland HD (1984) The chemical evolution of the atmosphere and oceans Wiley-Interscience, New York
Rollinson H (2007) Early Earth system. Blackwell
Windley BF (1975) The evolving continents, 3rd edn. Wiley

Hydrosphere

Berner EK, Berner RA (1978) The global water cycle: Geochemistry and environ-
    ment. Prentice Hall, Englewood Cliffs, NJ
Holland HD (1978) The chemistry of the atmosphere and oceans. Wiley
Millero FJ (1996) Chemical oceanography, 2nd edn. CRC

Biosphere and Soils

Birkeland PW (1999) Soils and geomorphology, 2nd edn. Oxford University Press,
    New York
Bolt GH, Bluggenwent MG (eds) (1978) Soil chemistry. Elsevier, Amsterdam
Jenny H (1980) The soil resource. Springer, New York
Lunine J (1999) Earth, evolution of a habitable world. Cambridge University Press,
    Cambridge
Schlesinger WH (1991) Biogeochemistry. An analysis of global change
Schopf JW (ed) Major events in the history of life. Jones and Bartlett, Boston
Schopf JW, Klein C (1992) The proterozoic biosphere. Cambridge University Press
Vernadsky VI (1997) The biosphere. Copernicus Books, New York

Geochemistry and Environmental Geochemistry

Chapman NA, McKinley IG, Hill MD (1987) The geological disposal of nuclear
    waste. Wiley, Chichester
Holland HD, Trekian KK (eds) (2004) Treatise on geochemistry. Elsevier
Mason B (1958) Principles of geochemistry, 2nd edn. Wiley
Shikazono N (1997) Chemistry of Earth system. University of Tokyo Press
    (in Japanese)
Shikazono N (2010) Environmental geochemistry of Earth system. University of
    Tokyo Press (in Japanese)

Geology

Holmes A (1978) Principles of physical geology, 3rd edn. Nelson
Siever R, Grotzinger J, Jordan TH (2003) Understanding Earth. Freeman,
    New York
Skinner BJ, Porter SC (1987) Physical geology. Wiley
Tarbude EJ, Lutgens FK (1998) Earth: an introduction to physical geology, 6th edn.
    Prentice Hall, Upper Saddle River, NJ

Geophysics

Holmes A (1978) Principles of physical geology, 3rd edn. Nelson
Lillie RL (1998) Whole Earth geophysics. Prentice Hall

Plate Tectonics and Plume Tectonics

Condie KC (1989) Plate tectonics and crustal evolution. Pergamon
Condie KC (2001) Mantle plumes and their record in Earth history. Cambridge
   University Press
McElhinny MW (1973) Paleomagnetism and plate tectonics. Cambridge University
   Press
Uyeda S (1978) New view of the Earth. Freeman

Atmosphere, Climate Change

Frakes LA (1979) Climates throughout geologic time. Elsevier, Amsterdam
Graedel TE, Crutzen P J (1993) Atmospheric change: an earth system perspective.
   Freeman, New York
Robison PJ, Henderson-Sellers A (1999) Contemporary climatology, 2nd edn.
   Longman, Harlow
Rowland FS, Isaksen ISA (eds.) (1988) The changing atmosphere. Wiley
Ruddiman WF (2001) Earth's climate: past and future. Freeman, New York

Planetary Sciences

Urey HC (1952) The planets. Yale University Press
Ringwood AE (1979) Origin of the Earth and Moon. Springer, New York

Environmental Science

Carson R (1962) Silent spring. Houghton Mifflin
Fuller B (1969) Operating manual for spaceship Earth. Southern Illinois University
   Press, Carbondale
Gore A (1992) Earth in balance: ecology and the human spirit. Houghton Mifflin,
   Boston
Lovelock JE (1979) Gaia. A new look at life on Earth. Oxford University Press
Meadows DH, Meadows DL, Renders J (1992) Beyond the limits. Green

Index